Reinventing Sustainability

How archaeology can save the planet

Erika Guttmann-Bond

OXBOW | books
Oxford & Philadelphia

Published in the United Kingdom in 2019 by
OXBOW BOOKS
The Old Music Hall, 106–108 Cowley Road, Oxford OX4 1JE

and in the United States by
OXBOW BOOKS
1950 Lawrence Road, Havertown, PA 19083

Paperback Edition: ISBN 978-1-78570-992-0
Digital Edition: ISBN 978-1-78570-993-7 (epub)

A CIP record for this book is available from the British Library

Library of Congress Control Number: 2018964660

Typeset in India by Versatile PreMedia Services. www.versatilepremedia.com

For a complete list of Oxbow titles, please contact:

UNITED KINGDOM
Oxbow Books
Telephone (01865) 241249, Fax (01865) 794449
Email: oxbow@oxbowbooks.com
www.oxbowbooks.com

UNITED STATES OF AMERICA
Oxbow Books
Telephone (800) 791-9354, Fax (610) 853-9146
Email: queries@casemateacademic.com
www.casemateacademic.com/oxbow

Oxbow Books is part of the Casemate Group

Front Cover: Terraced rice field near Sapa, north Vietnam. Photo by Blue Planet Studio, © Shutterstock.

This book is dedicated to two important men in my life:

To my husband,
Jerry Bond

And my brother,
Hans Guttmann

Contents

Abbreviations

ADE	Amazonian Dark Earth
FAO	Food and Agriculture Organization of the United Nations
GIS	Geographical Information System
ICARDA	International Center for Agricultural Research in the Dry Areas
IPCC	Intergovernmental Panel on Climate Change
IPM	Integrated Pest Management
NGO	Non-Governmental Organisation
OSL	Optically Stimulated Luminescence
TRIPS	Trade Related aspects of Intellectual Property rights
UNESCO	United Nations Educational, Scientific and Cultural Organization

Acknowledgements

This project was supported partly through crowdfunding, and I want to thank my donors, in particular Scott Weber and Britt Ellis for their generous contributions. I also want to thank Geoff Lee and Jules Pretty for commenting on the drafts – it was encouraging and very helpful to get that feedback. Many thanks to my supportive husband Jerry, who listens patiently. Thanks also to my brother Hans, who has unmitigated faith in this project, and to my father, Allen Guttmann, who also helped with funds. I would also like to thank my students, particularly the class who heard my first 'How archaeology can save the planet' lecture and honoured me with a standing ovation. Here's looking at you, kids.

Foreword

I have been thinking about this topic for about 35 years, inspired initially by a lecture that I had as an undergraduate. Mark Papworth was my favourite professor, and he told us many stories that I remember to this day – but best of all was his lecture about the archaeology of the Negev. Papworth gave us a blow by blow account of how a team of scientists discovered and interpreted huge landscapes of stone mounds and ridges in the Israeli desert, and how at first they thought that these features might be geological formations. Later, they realised that the stone structures were entirely man-made, and were associated with ancient farms in the middle of the desert – farms which flourished simply by the efficient collection and storage of rainwater. A team of scientists rebuilt one of these ancient farms, together with its water conservation system, and they found that it produced abundant crops even during severe droughts.

I left the class wondering why this extraordinary and sustainable system wasn't now the norm for desert farming, and years later I read a PhD thesis that explained precisely the problem. The PhD candidate wrote that the sustainable desert farming system in the Negev is extraordinarily productive and resilient in the face of droughts – but you can't cultivate it with a tractor, and therefore it is unfeasible. This myopic conclusion suggested to me that the problem was not financial or technological: it was simply due to a failure of the imagination.

In 2005 I gave a lecture on sustainability in the past to the agriculture department of a leading UK university, thinking (naively) that I could inspire the same enthusiasm that I felt as an undergraduate. Instead, they scoffed. One elderly professor said to me, 'It's a nice idea, *dear*, but it's pie in the sky.' I was taken aback by their attitude, and explained that what I was proposing is already happening in a number of countries, but they weren't prepared to listen. The next time I gave this talk I was better prepared: this time I had a powerpoint slide on which I listed about 15 countries that had already very successfully adopted the kinds of ancient technologies that I was advocating. Today, the list has become so long that I can't fit them onto one slide – I just say that on every continent, there are dozens of countries that are re-implementing ideas from the past. Then I give examples, one by one. People still scoff, but it is the successful farmers who are having the last laugh. Throughout this book I give examples of farmers in developing countries whose lives are being transformed by sustainable agriculture based on traditional knowledge. Many of these farmers are now turning profits that enable them to grow their businesses, increase their food security, and send their kids to school – which is helping whole communities. I want to see more of this: much more.

The Food and Agricultural Organization of the United Nations (FAO) is also advocating the use of traditional knowledge, having discovered that the most productive farms on earth are the small ones, particularly the ones that conserve soil and water. No one is suggesting that we reject modern technology, but my premise is that we can marry up old and new ideas and technologies in order to create more resilient farms and cities. We can conserve resources, increase production, lower costs, raise biodiversity and make our towns and countryside cleaner, greener and more liveable in the process. What's not to like?

I work at the crossover of the environmental sciences and archaeology, working to understand how people lived, farmed, and managed or adapted to their environments in the past. I began my studies in anthropology, but my great passion is – wait for it – soil science. To most people, soil (please don't call it 'dirt'!) is simply something to be washed off their children's clothes, but when you look at it closely, soil is amazing. There are a host of microbes and fungi performing small miracles that enable plants to grow and organic matter to decay, while chemical and physical processes recycle the nutrients that are needed by all living things. As a colleague of mine once said, 'Soils are great! You got *inputs*, you got *outputs*, and in between you got PROCESSES!'

Under certain conditions, soils and sediments can also hold preserved pollen, which provides us with great detail about the trees, crops and other plants that covered the landscapes of the past, while also showing us how the vegetation changed over time. For my MSc I analysed the fossil pollen of a coastal landscape in Cornwall, all the while picturing so vividly the plants and ecosystems I was identifying that one night I dreamed I was standing on a hill looking out at a verdant green landscape made out of my own data. The grassy hills were formed by bar charts, with each bar representing levels of different mineral components in the solid geology, while representations of the different soil strata cut horizontally across the bars, each layer cross-hatched with the drawing conventions that tell scientists the characteristics and origin of each layer. Around me there floated grains of pollen the size of golf balls, representing pictorially all the vegetation I could see around me. It was an environmentalist's heaven.

Archaeology is my other love. I was drawn to archaeology because I always wanted to be an explorer, and the past is a country – or rather many different countries – that will never be fully mapped. I also like being outdoors, and I like bashing things with a mattock, but archaeology allows me not just to be physically active: it allows me to think and to analyse, to try and interpret what I am digging and to try and piece together ancient landscapes. I like the process of looking for clues: looking at the lumps and bumps in a field, looking at old maps, noticing the old greenways and identifying the Roman roads that cut through the earlier landscape, or observing the relics of the open field systems that characterised the Middle Ages in Britain.

I work mainly in the UK, having lived here since 1989, but I have also lived in America, Germany and the Netherlands, and I've worked in Israel, Jordan, Egypt, Greece and Ireland. I have mainly taught in the UK, but I also taught for two years in the Netherlands, as Professor of Landscape Archaeology. This enabled me to study a landscape that has some parallels with the UK – which I will discuss in the chapters that follow.

The scope of this book is huge, because I'm talking about a concept that is global in scale. One reviewer said that this book should not be written, because it's too large a topic, but I disagree. Natural Historians once said that ecology is too big a concept to teach, and argued that botany, biology and zoology should all be taught separately. The professor who suggested teaching them together was actually laughed out of the room by his ungracious colleagues, but now it is recognised that you can only study complex systems by looking at how the different components interact.

I have divided this book into chapters that represent different kinds of environments, so that I can introduce the different ways that people have adapted to each of these situations and the particular opportunities that they offer. Chapter 1 provides an introduction to the scope of the book, also providing the reader with some information on the methods used by archaeologists and environmental archaeologists. I am assuming that the reader has curiosity but no specialist knowledge. In Chapter 2 I discuss how people have adapted to living in wetlands, including how farmers have learned to turn the disadvantages of swampland into advantages. As global sea levels rise, many governments (including the UK) are looking at a system of 'managed retreat', in which we turn some of our coastal land back into wetlands. This does not have to mean that we give up the land completely, and I give examples of different ways that people have successfully lived in wetlands, and even farmed the wetlands.

In Chapter 3 I discuss deserts, with a focus on rainwater catchment and storage systems that were developed in the past. As the climate warms and the deserts spread, water is becoming increasingly scarce, but ancient rainwater collection systems are now being adopted to combat this problem. Chapter 4 discusses ways of enhancing food security by returning to traditional methods such as intercropping (also known as polycropping), including a number of recent success stories from around the world. In Chapter 5 I discuss soil degradation and erosion, and the many ways that we can fight these evils. Soils can be created, as well as destroyed, and sustainable agriculture can be introduced into developing countries with no technology or limited technology, but also to fully industrialised nations such as those in North America. South American farmers are now leading the way in adopting conservation tillage, which is saving modern farmers a fortune while at the same time slashing the use of fossil fuels.

In Chapter 6 I discuss vernacular architecture, and the ways that traditional architecture is adapted to local conditions and environmental extremes. There are cheap and effective ways of building in the desert, and traditional architecture in many different countries shows us how to keep interiors cool in the midday sun and warm in the cold desert nights. Vernacular architecture can also provide us with a model for development in geologically unstable regions, where traditional architecture is often built to withstand the stresses of earthquakes. Chapter 6 also discusses green spaces in towns in the past; greening up our cities with urban gardens is providing leisure and improved nutrition to city dwellers, while at the same time absorbing pollutants and reducing food miles.

Chapter 7 brings it all together. Here, I sum up not just the ways we can put traditional knowledge back to work in the developing and developed worlds, but I also discuss the extraordinary new technologies that can make the implementation both efficient and cost effective. I discuss green energy, new methods of construction and conservation, and the new discoveries that reinforce what we already know intuitively: that green spaces are important not just for our physical health, but also for our emotional well-being.

None of the conservation methods I discuss here are 'pie in the sky'. Throughout this book, I give examples of how traditional knowledge is already working to save the planet. I want to see these developments snowball. My students have gone out and performed their own experiments, one recreating the ultra-productive and carbon-storing Amazonian Dark Earth soil here in west Wales, and another introducing the perfect ecological circle known as the 'Three Sisters' intercropping system to an organic farm in the Netherlands. I hope that this book sparks off many more new projects, and at the very least I hope it gives people some cause for optimism in today's anxious world.

1

Learning from the past

There have been many books written about what we can learn from the failures of the past, and there are many examples of ecological disasters that came about because of our over-exploitation of the environment.[1] I want to take a more optimistic view, focusing on what we have to learn from past *successes*. This book is about sustainable agriculture in the past, and the engineering works that supported it. It is about traditional agricultural methods, some of which survived industrialisation and continue to be used in developing countries. It is also about sustainable building traditions that are adapted to local conditions and climatic extremes. Finally, it is about taking ancient ideas, integrating them with modern scientific observations, and putting them back to work in ways that can revitalise our towns, our countryside and our economies. This is already happening in countries all around the world today. We have already experienced an industrial and a technological revolution; now, we are going through a revolution in the way that we think.

There have been considerable changes to the countryside of the developed world in the last 200 years, changes that have accelerated throughout the 20th century. In Britain, hedges have been removed in order to make larger fields for crops that can be mechanically planted, fertilised and harvested. Biodiversity in industrialised agricultural landscapes everywhere has plummeted because large fields have been planted with single crops (or 'monocrops') and sprayed with weed-killers to ensure that only one species (the crop itself) survives. Machines have taken over most of the manual labour on farms, changing the nature of rural life and vastly diminishing the number of jobs in the countryside. Animals are farmed more intensively, with higher stocking densities, and are taken greater distances to abattoirs and markets.

Agriculture is a vitally important aspect of our past, present and future, but farmers do a lot more for us than just growing our food: it is the farmers who manage our countryside. In many places the countryside is regarded as an amenity that is there for everyone to enjoy, but if we want to continue to enjoy it, we need to try and preserve the landscape and the remaining native plants and animals that have managed to withstand the deprivations of industrialised farming.

The methods used in *indigenous* or *traditional* agriculture (by which I mean pre-industrial agriculture) developed over thousands of years, through the process of trial and error.[2] This means that they can be regarded as long term experiments, in which unsuccessful methods were either abandoned or modified. Many of these ancient methods have been lost or forgotten, but are now being re-discovered by archaeologists, and experiments in recreating the old systems have demonstrated that they can be more sustainable *and* more productive than the most modern farms in the same regions.[3] Obviously there are other obstacles – political instability, war, issues of land ownership and other economic considerations – but these issues have been covered elsewhere; this book aims simply to demonstrate that we already have the technology to make substantial improvements in the way we manage the countryside, both in the developed West and beyond.

I will also argue for the importance of vernacular architecture, *i.e.* traditional building techniques using local materials. Local styles reflect local cultural and artistic traditions, but are also the result of centuries of trial and error. I am particularly interested in pre-industrial methods of heating and cooling interiors, and of building styles that are resistant to local climatic extremes. I'm also interested in vernacular approaches to building earthquake-resistant housing. Concrete, glass and steel have their place, but concrete buildings can be hot in summer, cold in winter, and prone to collapse in earthquake zones, so they do not always represent the most logical style of building.

In Europe we are seeing a revival in vernacular architecture in some areas, partly because of the sustainability and eco-friendliness of using local materials such as earth or mud brick, and partly because of the beauty and distinctiveness of traditional buildings. These structures can readily incorporate centuries-old building techniques together with modern conveniences such as solar panels, geological heat pumps and other modern adaptations – and can do so without costing the earth.

The evidence that I present in this book draws on the fields of archaeology, anthropology and history. The scope of this book is global, and the time span is largely focused on the last *c.* 11,000 years, since the origins of farming. By taking this broad approach, we can see that the same solution to any particular environmental condition was sometimes discovered independently by different people at different times and places; agriculture, for instance, was developed independently at different times and on different continents. Another example is water harvesting, which allowed people to farm the deserts of North Africa, the Middle East and the Americas.

The ambitious scope of the book is necessary because I am dealing with concepts that transcend particular times and places; I am looking at the many ways that people have adapted to often very different environmental conditions, and I want to make the case that many of these practical solutions are still applicable today. We can reinvent the ancient and historical techniques in a new age, and in doing so we can integrate them with new technologies such as renewable energy, drip irrigation, satellite data collection and information-sharing on a global scale.

Intermediate or appropriate technology

The Green Revolution was a well-intentioned movement that took place in the 1960s and 70s, which attempted to solve the world food crisis by bringing modern industrial farming to developing countries. Inorganic fertilisers, pesticides and machinery had revolutionised agricultural development in the West, and it was hoped that these innovations could benefit countries that were still practicing traditional forms of agriculture. The new introductions resulted in increased outputs initially, but over the years productivity levelled off or declined, and the environmental damage was, in many places, substantial. Rachel Carson described the ravages of industrial farming in the West in her classic book, *Silent Spring*, which provoked a widespread awareness of the dangers of pesticides and other pollutants.[4] Heightened awareness led to better legislation to protect the countryside in the West, but developing countries do not have the same degree of regulation, and there have been many health problems as a result. A further problem is that developing countries often do not have the facilities to operate or repair the machines that they were given; in the words of Ernst Schumacher:

> In every 'developing country', one can find industrial estates set up in rural areas, where high-grade modern equipment is standing idle most of the time because of a lack of organisation, finance, raw material supplies, transport, marketing facilities and the like.[5]

Schumacher noted that what was needed in developing countries was simple technology that could be easily built and maintained by local people using local materials. He called this 'Intermediate Technology', and he set up an organisation (today called Practical Action) which works to introduce or advance such technologies in developing countries around the world. The idea is that poor countries tend to have sufficient labour but insufficient capital, and they benefit from technologies that are more labour intensive but less expensive in terms of raw materials and maintenance. The term 'appropriate technology' is now used to mean more or less the same thing. Schumacher was drawing on the ideas of Mahatma Gandhi, who objected to technology that takes control away from ordinary people. Gandhi advocated the use of simple machines such as the sewing machine (which can be powered by pedal) and the bicycle – both of which can significantly improve the lives of poor people living in villages without electricity.

Many ancient technologies are defined by engineers as 'intermediate': they are simple, low in cost and they depend on local materials. Significantly, they do not require fossil fuels. There is now broad agreement among many governments, non-government organisations (NGOs), engineers and agronomists, as well as the United Nations, that intermediate technologies are often the most appropriate way forward in developing countries. This is not an attempt to deprive the needy of modern technology, but rather it is the recognition that it is often more effective to use the technology that they already possess – or that their ancestors possessed.

In the 19th century West, modern technology was hailed as the cure for all human ills, and it was seen as the way forward to a better life. Technology was equated with 'progress' in a world of linear time, with a beginning (the Creation) and an end (the Day of Reckoning). The natural environment was regarded as a source of material wealth and was mined, cleared and built upon accordingly. By contrast, in many non-Western societies time has a different configuration, with all things working in cycles.[6] This is perhaps a more helpful way of looking at the world, as sustainability is a key element: we cannot afford to use up all our resources, because (all things being well) the world is not going to end any time soon.

While modern technology is moving forward in leaps and bounds, at the same time we are losing vast amounts of traditional knowledge and technology. One engineer I spoke to at Cambridge University said he suspects that we may actually be losing more information than we are gaining. Rather than discarding the knowledge and values of the past, surely we can find a way to develop new technologies that complement the traditional ways of life? The sophistication of early farmers has often been overlooked or underestimated by modern researchers because we have failed to recognise that farmers in the past had different priorities to farmers in modern, developed countries. Subsistence farmers – as opposed to agribusinesses – depend absolutely on their crops, and their emphasis is therefore on the security and sustainability of their agricultural systems, rather than the productivity per hectare (although production of a surplus was of course welcome, and provided food security, opportunities for trade and also a way to enhance social ties through gift exchange). Many ancient agricultural systems are in fact secure, sustainable *and* productive, and studies taking place all over the world today are showing that properly implemented sustainable agriculture actually tends to produce *higher* crop yields than crops raised using conventional industrial methods.[7]

I will be discussing all these issues further on, but I will include an example here, to illustrate my point. In Bangladesh, modern, high yielding seeds, pesticides and inorganic fertilisers were introduced in the 1960s, and production went up – initially. In the next 30 years, yields began to decline, although more and more fertiliser was added every year. To make matters worse, the price of fertiliser tripled in that time. People's health was suffering, miscarriages were increasing, animal health was in decline, the soils hardened as soil organic matter dwindled, and the fish died as water quality deteriorated.[8] Farmers all across the country went bankrupt, and many moved to the cities. Then, flooding in 1998 destroyed the livelihoods of many of the surviving farmers.

A local NGO[9] worked with the farmers in one of the worst hit flooding areas. They suggested that the farmers go organic, an idea which was initially regarded with suspicion by the men. They wanted a modern solution, and did not want to be fobbed off with something out of the past. Women, on the other hand, were interested, and gradually the idea caught on as locals became more aware of the human and environmental costs of industrialisation that they had experienced in the recent past.

The locals call the new system 'Nayakrishi Andolon', the New Agriculture, to show that it isn't just a return to a pre-industrial past – rather, this is something that combines traditional agriculture with new ideas. While farmers of the Green Revolution had produced monocrops of sugar cane, farmers of the New Agriculture grow seven different crops together, as they did in the pre-industrial past. Growing crops together, or 'intercropping', is far more productive per hectare than growing single crops, and it makes the crops less susceptible to pests and diseases. In the new/old system, legumes are being planted to put nitrogen back into the soil (see text box), and compost is made out of water hyacinth, which had been regarded as an intrusive weed. Banana leaves, rice straw and cow dung are also composted and added to the soil, and as a consequence the ground has now softened, worms are active again and the soil has once again become rich in organic matter. Sixty-five thousand Bangladeshi households have now gone over to the New Agriculture, and the farmers themselves are spreading the word.[10] By embracing sustainable agriculture, the Bangladeshis have recovered a link with their past, they have improved yields, recovered their health, and they have revitalised their economy.

Nitrogen fixation

Nitrogen-fixing plants have bacteria in their roots, which take up nitrogen from the air and transform it into a form that is available to plants. Legumes are the most important biological provider of soil nitrogen. Nitrogen-fixing plants provide energy in the form of carbohydrates for the bacteria, and the bacteria provide nitrogen for the plant. This kind of mutual assistance in the natural world used to be called 'symbiosis', though it is now generally called 'symbiotic mutualism'.

What is sustainable agriculture?

Sustainable agriculture is farming that does not pollute the environment, and does not deplete the natural resources on which it depends. It aims to conserve the resources that are available on a farm, to maximise the 'natural capital' that already exists, without diminishing it. This means conserving soil by protecting it from erosion, and conserving and enhancing the nutrients within the soil through the addition of organic waste produced on the farm, or by planting nitrogen-fixing crops. The insects that eat crop pests are encouraged, and the habitats that these helpful predators depend on (such as hedgerows) are maintained. Sustainable agriculture mimics natural ecosystems, which is why it is also known as 'agroecology'.

Modern, industrial agriculture relies heavily on pesticides, herbicides, inorganic fertilisers, animal feeds and fossil fuels. It relies on 'external inputs', materials brought in from off the farm. In a sustainable system, the farmer imports little or no material from off the farm, but uses natural systems to mimic the actions of pesticides and fertilisers.

Traditional farms used to maintain both animals and crops, because crops could not be grown very successfully without the nutrients and organic matter from the animal manure. Organic material retains water, which is especially important in sandy soils that drain too quickly, and it adds nutrients to the soil. It provides food for earthworms, which are essential to soil health: they aerate the soil, increase soil stability and cohesiveness, and they enhance the availability of nutrients to plants. Simply adding inorganic fertilisers and no compost or manure will lead to depletion of the soil organic matter, making the soil less cohesive, less biologically active and more readily eroded. Without the organic matter and the microbes and worms that depend on it, soil can become compacted or it can simply turn to dust and blow away.

Soil organic matter is the critical element in the agriculture of arid regions, for example in Senegal. This West African country lies within the Sahel, a band of semi-arid grassland where the soils are sandy and desertification is a serious problem. Adding inorganic fertiliser and pesticide is too expensive for most of the local population, but these additives don't really help much anyway.[11] Because there is so little organic matter in the sandy soil, there is nothing to bind the nutrients, and the fertiliser simply washes away in the rain or is taken up by the weeds and soil microbes.

The Rodale Institute, an organic agriculture research organisation, has helped introduce sustainable systems to Senegal that are really making a difference. Local farmers have integrated livestock into their farms, so they can improve the soil with manures, and they have begun growing legumes, which add nitrogen to the soil. Composting and water harvesting have also helped, and together these sustainable improvements have pushed up yields by between 75% and 195%.[12] A study commissioned in 2005 on a number of sustainability projects all across Africa demonstrated that introducing sustainable systems – including soil conservation, integration of livestock and arable, integrated pest management and many others – had succeeded in more than doubling yields in three to ten years.[13] We are now seeing a growth in sustainable agricultural systems around the world, and I believe that sustainable intensification – growing more crops on the same piece of land, without depleting that land – has got to be the way forward.

Is sustainable agriculture the same thing as organic?

There are degrees of sustainability, and many would argue that the only way to produce sustainable crops is to go completely organic. Others argue that inorganic fertilisers can be part of a sustainable ecosystem if they are used sparingly. The proponents of inorganic fertiliser point out that if you apply too much of any fertiliser – either organic or inorganic – the excess will leach out of the soil and pollute the groundwater. Introducing or improving sustainability is not an 'all or nothing' approach, and all kinds of small steps can be taken that will save money as well as saving natural resources. There have been a number of studies that demonstrate that fertilisers can be applied a lot more rationally, saving money and reducing groundwater pollution.[14] Soils can be tested and fertilisers applied to the parts of the fields that need it most, instead of applying the same amount to the entire field. Insecticides can also be applied in a more targeted way. In the 1980s it was discovered that spraying rice crops in Asia actually *increased* numbers of the key crop pest, the brown plant hopper, because the spray killed off the insects that eat the plant hopper.[15] They have not given up spraying altogether, but have found that the most efficient method seems to be to monitor the situation and spray only when the plant hopper population grows out of control.

In Denmark the use of fungicide has been substantially reduced by placing weather stations on the farms, which help farmers predict the conditions that lead to outbreaks of disease.[16] This means that they can delay applying fungicide to their potato crops for up to five weeks or more, instead of wasting the spray by applying it too early in the season, and causing needless pollution.

Another compromise is a reduction in ploughing. Ploughs break up the soil to prepare a seed bed, breaking down heavy clays and allowing in the oxygen that the crops need to grow. In the Middle Ages the mouldboard plough was invented, which turns the soil over completely and so has the added benefit of killing off annual weeds. In fact, by looking at the charred seeds that survive on archaeological sites, we can see that the whole suite of agricultural weeds changes at this time.[17] Ploughing also has its drawbacks, however, one of which is that tractors compact the soil so that water runs off instead of sinking in, causing erosion and gullying in the fields. Ploughing also exposes the organic material in the soil to the air, so that it decays and releases its carbon; this makes the soil less fertile, and makes it less able to hold water and nutrients, and it releases carbon dioxide into the atmosphere. Reduced tillage and 'No-till' agriculture have been suggested as a way forward, and in response to this there has already been a 35% reduction of ploughing in the US. By 2005, 23% of America's arable land was being farmed by no-till,[18] and reduced tillage was practiced on around 41% of American farmland.[19] It is even more widespread in South America, with 65% of farmland in Paraguay under a no-till regime by 2005.[20] By 2013, 157 million hectares of cropland worldwide – 11% of the world's arable land – was under no-till or low-till regimes;[21] by 2017, that figure had risen to 180 million hectares.[22]

A recent survey estimated that farmers are applying 3.5 billion kilograms of pesticides per year worldwide, but it has also been estimated that more than half of this pesticide is not actually necessary.[23] Recent studies show that 29 Midwestern American cities have herbicides in the tap water, while the average American has detectable levels of 13 different pesticides in their body.[24] We can do better than this. Integrated Pest Management, or 'IPM', is the science of natural pest control. This group of systems relies on naturally occurring pesticides such as those occurring in the leaves of neem trees (*Azadirachta indica*), using pheromone pest traps, the use of predators to attack insect pests, rotating crops and other methods. Turning again to Bangladesh as an example, IPM has raised yields by 40–130% and farmers' income has risen by 300% due to increased production and lower production costs.[25]

There are many traditional, low-tech systems that survived the Green Revolution in developing countries, but new research in sustainable agriculture can be used to supplement these old and established methods. I am not advocating a wholesale return to past technologies, nor am I suggesting the adoption of early technology in place of modern engineering and agriculture. What I am suggesting is that we can use some aspects of early technology, and we can combine old, tested techniques with new systems and inventions, to create a healthier, more sustainable and environmentally richer planet.

How do we study agriculture in the past?

Archaeology

Archaeology is about a lot more than artefacts, and in modern practice it incorporates a great range of different scientific disciplines. One important area of research is

Figure 1.1: A colleague recording samples from the layers in a vertical section face at Sde Dov, Tel Aviv. (Photo © the author)

Figure 1.2: The author planning a Roman site in the Conwy valley; the planning frame and graph paper enable me to draw at a scale of 1:20. (Photo © Jerry Bond)

Environmental Archaeology, which encompasses subjects like botany, entomology, zoology and geology to help us understand how people interacted with their environment in the past. But first, a word about archaeology and how it works.

Archaeology developed out of an interest in the artefacts of the past, and the early archaeologists, known as 'Antiquarians', more or less ransacked important sites in order to obtain the 'finds'. One of the first people to reject this attitude was Thomas Jefferson, who later became the third president of America. Jefferson oversaw the excavation of a burial mound on his property in Virginia – an activity popular with the landed gentry of England, but whereas most people were simply plundering, Jefferson approached it with the aim of recovering information as to its origins. He recorded the different layers that made up the mound, drawing the section (the vertical slice through the layers – see Figure 1.1) to show the many phases of use and re-use of the mound where the dead were buried over the course of many generations.

This rational approach did not catch on at the time, and it was not until the 19th century that archaeologists such as the British General Pitt-Rivers (1827–1900) began organising excavations where drawing plans and sections became standardised (Figure 1.2). Pitt-Rivers saw to it that ALL artefacts were recovered, not just the more exciting ones, and he initiated a trend that led to archaeology becoming the pursuit of information, rather than simply the pursuit of treasures. Sir Mortimer Wheeler continued in this vein, recording the stratigraphy (which means noting which layer came first), and keeping track of the chronological order of deposits. Wheeler introduced a methodical grid system in which baulks were left between excavated areas so that layers could be understood and recorded more efficiently.

Many people continued to pillage sites rather than excavating them scientifically, but over time the field developed and – in many countries – archaeology came to be practiced in a methodical manner. In practice, this means recording each distinct layer, noting its colour and stoniness, working out whether it accumulated rapidly or slowly, through natural sedimentation processes or by man-made activities. Archaeologists can not only trace ancient layers and walls, but we can also detect the different sedimentary deposits that show the location of the *cut* of each feature, for instance the cut of the trench that was dug in order to build a wall foundation. Every activity that can be identified is recorded, so that the sequence of events on a site can be understood. We can see when walls were built and destroyed, and when the masonry was finally removed for re-use elsewhere, leaving behind ragged looking 'robber trenches'. We try to interpret every piece of information to build up a picture of how the site developed and grew, and eventually, how it was buried.

European Chronology

Mesolithic: 9500–7000 BC (hunter gatherers)
Neolithic: 7000–2000 BC (first farmers)
Bronze Age: 2000–800 BC
Iron Age: 800–50 BC
Roman: 50 BC to AD 476, ending with the fall of the Western Roman Empire.
Byzantine: AD 476–AD 1453, ending with the fall of Constantinople.

By establishing the stratigraphy of a site – the chronological order in which things took place – we can also establish a relative chronology for the artefacts. In a very broad way, this was done by a 19th century Danish archaeologist, C. J. Thompsen, who coined the terms 'Stone Age', 'Bronze Age' and 'Iron Age'. The earliest artefacts we discover are of stone and flint (and also organic materials, but these do not survive very often). The discovery of combining copper and tin to make bronze occurred around 3000 BC, although no one is sure exactly where this technology was discovered. It is known in the Near East, Egypt, parts of Asia and in the Aegean, but it did not reach central Europe until about a thousand years later. Finally, iron was developed in the Near East around the 12th century BC, reaching Central Europe by the 8th century BC. We still use the Three Age System, as it is called, but there are other aspects of cultural change that are more important than the changing technology – for instance, the transition from hunting and gathering in the Mesolithic (middle Stone Age) to farming in the Neolithic (the new or latest Stone Age), the changes in burial customs and land management in the Bronze Age, and the origins of towns and the development of more complex, hierarchical societies in the Iron Age.

The sequencing of artefact types was established on many sites, and correlations could be made from site to site where similar artefacts were uncovered. These classifications enabled archaeologists to establish a system of relative dating, so it was possible to say which artefact type came first, but not so easy to say exactly when it was made. In the historical periods there were texts to assist in the chronology, and the production of coins enabled absolute dating to take place, so for instance when you find a particular type of pottery together with coins of a certain date, you have a general idea of when the pottery was made.

Dendrochronology, or tree-ring dating, was developed in the early 20th century. The principle is based on the fact that most trees have growth rings that vary in size, depending on the climate during each season and also the species and age of the tree. Because trees form a distinctive ring pattern in response to the climate, a segment of tree rings produced by a tree that grew between AD 1700 and 2000 will have an overlapping span of rings that will be identical to a nearby tree of the same species that was growing between 1600–1800. By looking at overlapping patterns of increasingly older timbers, a master chronology is built up for the region. The master chronologies for various regions depend on living trees for recent years and use timbers preserved in medieval and later buildings for the historical period. Timber preserved in wetlands enables us to carry the chronology back into prehistory, and at the moment, the chronology for the Northern Hemisphere goes back about 13,900 years.[26] Dendrochronology and the link between tree rings and climate have been important in helping climate scientists to identify climate change in the past.

It was not until radiocarbon dating was developed by Willard Libby in 1949 that absolute dating could take place on a regular basis. Radiocarbon dating does not provide exact dates, but it provides a date range which enables us to roughly work out the timing and duration of archaeological events. This is important for understanding the development and timing of changes in culture, and in people's changing interactions with their environment.

Radiocarbon dating

Radiocarbon (^{14}C), is an unstable isotope with 6 protons and 8 neutrons. It decays at a constant rate, slowly turning into Nitrogen 14. Since all living things contain carbon, they therefore take up radiocarbon from the atmosphere. This means that anything organic (*e.g.* charcoal, bone and shell) can be measured and dated.

Environmental Archaeology

Archaeology has always borrowed from the sciences, particularly geology, but in the 1960s and 70s this trend increased exponentially, leading to the development of the field of Environmental Archaeology. This is a broad field of study which includes archaeobotany (the study of ancient seeds, pollen and other plant remains), zooarchaeology (the study of ancient animals, both vertebrates and invertebrates), and geoarchaeology (including geology, soils and sediments).

Ancient seeds can be preserved by charring, and are recovered from excavations by placing soil from archaeological deposits in water, which makes the charcoal rise to the surface. Using a low powered microscope, the seeds can then be identified to family, genus or even species, which tells us not only what people ate, but also what sorts of weeds grew in their fields. Through archaeobotany, we have discovered that people in the past grew different wheat varieties together in one field; through anthropology, we can understand why they did so. Growing different crop varieties at once is a way of ensuring a crop every year, because different varieties are resistant to different environmental stressors: in a dry year, the drought resistant strains will thrive, whereas in a wet year, the varieties that are resistant to mildew and fungal attack will survive. There are thousands of different varieties that have been bred over thousands of years to resist different pests, viruses, drought, extreme cold, high winds and a range of diseases.

Archaeobotanists also study the by-products of cultivation, including the weed seeds that are collected together with the crops. Both cereal grains and the weed seeds that accompany them can be accidentally burned when grains are dried in primitive drying kilns, and these ancient drying accidents provide archaeobotanists with samples that provide insights into the ancient soils. The weed seeds tell us about the fertility of the fields, because different species or families have particular ecological niches and nutrient requirements. By looking at the weed seeds from a range of archaeological sites in any given region, we can see when people started farming the light, rich soils that were best for agriculture, and we can also see when they began to cultivate the heavier, more difficult farmland to which people had to resort when populations expanded and the best land became scarce. The weed seeds show us how farmers amended these difficult soils, making them viable for crops by breaking them up and adding organic manures. Weed seeds can also tell us about the extent of disturbance to the cultivated soil, which enables us to determine how intensively people in the past were ploughing; this helps us to understand levels of soil conservation and erosion in the past.

Pollen grains preserved in the oxygen-free environment of lake beds and peat bogs can tell us about early vegetation regimes, and palynologists (pollen specialists) have traced the changes in vegetation that have occurred over thousands of years. Pollen grains of cold-tolerant species begin to appear soon after the retreat of the glaciers 12,000 years ago, and we can see that wave after wave of new species colonised the newly exposed earth. Soil developed gradually, and was colonised by a sequence of herbs, shrubs, birch trees and eventually forest. Pollen analysis enables us to see when people arrive on the scene, when great gaps start appearing in the woodland, together with signs of human activity such as artefacts and fires. In time, we see indications of agriculture, when the woodland pollen declines and the pollen of cereals and arable weeds begins to increase. These are just a few examples; pollen can tell us about a number of different ecological changes that took place in past environments, and by investigating all the different strands of evidence we can put together people's place in the sequence. People can act as instigators, or they can react to change, or they can intensify changes that are already happening due to natural causes.

Animal bones tell us when wild animals first became domesticated, and zooarchaeology provides information about the vast range of interactions of people and animals. It is easy to forget that domesticated animals were not just kept for meat and milk, but also for wool, leather, manure for the crops, traction for the ploughs, transportation and – as we see in cultures all over the world – company. Animals have been kept simply as pets by people who would go hungry themselves rather than sacrifice what they regarded as a member of the family.[27] Zooarchaeology (a.k.a. archaeozoology) helps us to understand the domestication process, including how people changed the appearance and the behaviour of animals, creating breeds that were more easily managed and which were often well-adapted to local climate and conditions. Many ancient breeds have been lost, but the adaptive success and the hardiness of many early breeds is now recognised, and some farmers are working to keep these lines alive.

Early agriculture is also traced through geoarchaeology, the study of ancient soils and sediments. It is worth noting here that there's a difference between a soil and a sediment. Soils develop *in situ*, which means in place: layers of rock are slowly broken down into sand, silt and clay, which provide nutrients for plants. Plants die and decompose, adding organic matter to the soil. Earthworms bind the mineral and organic components together into aggregates, and the dung and decaying bodies of animals are incorporated into the mix.

One distinctive characteristic of a soil is that it forms naturally into layers, called horizons. Soil horizons develop through weathering processes, both physical and chemical. Frost, erosion and glaciation are physical processes that have a major effect on the soil. Chemical weathering acts to break down various elements within the soil; rainwater is slightly acidic and it interacts with the iron, organic compounds, clays and other soil constituents. These are then washed or leached out of one horizon and precipitated out in another layer below. Fluctuating water tables cause further changes, due to the presence or absence of oxygen, which interacts with iron in the soil. In hot countries, evaporation rather than rainfall is the most important process, and water and soluble elements in the soil are drawn upwards instead of washed downwards by the rain. Salt, in particular, can cause problems for farmers in arid regions because of this tendency to be drawn upwards towards the surface.

Sediments are distinct from soils because sediments are made up of material that has been *transported*. Rocks break down into sands, silts and clays, which become sediments when they are transported by wind and water. Eroded soil also becomes sediment once it has been moved from its original position. Soil on hill slopes that is washed downhill is called *colluvium*; if it then washes into a river it becomes waterborne *alluvium*, which is deposited in river deltas or floodplains. Stronger wind and water pressure transport larger-grained sediments, while slower moving water carries and deposits only finer sediments, so the particle size of a sediment can tell us about the environment in which it was transported and deposited. Particle size is therefore an important indicator for past environmental processes.

Deposits of eroded soil are an important source of evidence. Erosion happens naturally, as a result of wind, water and gravity, but when vegetation is removed, it makes the land much more vulnerable. There are natural causes for vegetation change, including climate change, but apart from glaciation, humans have had the most serious impact. People started removing trees to create farmland in Europe around 7000 years ago, and we can identify the consequences by the accumulation of eroded sediments beginning around this time. Periods of intensified agriculture can be identified by increased erosion, but we can also identify ancient and historic practices that would have checked erosion, for instance having smaller field sizes, terracing the hillsides, and ploughing with the contours.

Both soils and sediments can be buried by later soils and sediments, and this can more or less stop the processes of physical and chemical weathering on the buried layer. 'Buried soils' are essentially fossilised, retaining the chemical signature that tells us about their fertility. Since soils reflect the weather and climate regime of the place where they were formed, buried soils can act as a record for climates of the past, even thousands of years after they were buried. They can also hold records of past vegetation, because woodland, grassland, moors and other ecosystems are linked with particular types of soil. In woodland, the tree roots reach deep into the soil to recycle nutrients, breaking up the bedrock and forming a deep, rather uniform looking soil which often becomes more clay-rich towards the bottom. Thin soils on chalk can be linked with grassland, and moorland soils have distinct light and dark horizons where humic compounds, iron, manganese and aluminium have leached out of the upper horizon and accumulated in lower horizons, forming a distinct, colourful, stripy effect (Figure 1.3).

Figure 1.3: A distinctive type of soil called a 'podzol'. The white layer has lost its iron and aluminium, which have accumulated in the dark orange iron-rich layer below. (Photo © Randy Schaetzl, Michigan State University, Department of Geography)

Buried soils can retain their chemical signature as well as their physical structure, and this is important for the study of past agriculture. Plants take up nutrients from the soil, particularly nitrogen, phosphorus and potassium, but also a range of micronutrients including calcium, magnesium and sulphur, which perform a number of functions essential for plant growth. When plants decay, they give nutrients back to the soil, or if they're eaten by animals, the animal dung (and eventually the animal itself) goes back into the soil. Obviously, then, if the plants are harvested and organic material is not replaced, the soil quality will deteriorate. Using geoarchaeological methods, we can distinguish a good ancient agricultural system from an unsustainable one, by looking at things like nitrogen, phosphorus, soil structure and the quantity of organic matter in the soil.

Experimental archaeology

Another way to investigate early agriculture is by experiment. Archaeologists have re-created many of the features that we find in our excavations, in order to better understand the processes that created the remains. Buildings are reconstructed to see if we have understood them correctly, and various types of early industry have been attempted, including metalworking, glassmaking and ceramic technology. Experiments can be exploratory, or they can draw upon more formal, hypothesis-based testing.

Early farming techniques are a major area of interest, and experimental farms have provided a great deal of useful information. Simple wooden ploughs called 'ards' have been found preserved in waterlogged contexts, and ancient rock carvings show us how they were used. Three kinds of prehistoric ards can be identified; the rip ard has a share set at a steep angle, similar to the simple ploughs still used in Spain.[28] In Galicia, where it survives, it is used to break up new ground. It can be dragged along for 2–4 m before it drives itself into the ground and stops – which is in fact the same length as many of the ancient plough marks we see archaeologically.

Another type of ard was found at Donnerupland, Denmark (Figure 1.4). This plough was broken and worn when it was pulled out of a bog ('probably dumped there with a curse rather than a blessing', as the great experimental archaeologist Peter Reynolds dryly noted). It has a wear pattern that shows the ard was pulled along at a 29 degree angle, and experiments show that if pulled in this way, it doesn't bounce around or dig itself into the soil. A third type of ard may have functioned as a seed drill, cutting a furrow in the broken soil to plant the seed. Since 95% of the seed germinates when you sow it into a prepared drill like this, it takes more work but is more efficient than simply broadcasting the seed, which germinates only 25%. A feature that all these early ploughs share is that they did not actually turn over the soil. This means that prehistoric ploughing is a form of 'low till' agriculture that preserves organic matter in the soil. Whether or not farmers realised it, they were practicing what we now call 'conservation tillage'.

Archaeobotanists have identified domesticated wheat and barley dating back to the Neolithic, together with legumes such as lentils and peas. At Butser Experimental Farm, where Iron Age crops are grown using Iron Age methods, they tried rotating

Figure 1.4: A reconstructed ard plough, similar to the Donnerupland plough. (Photo © Jerry Bond)

legumes with cereal crops, which resulted in 2.6 tonnes of cereal per hectare compared to 1.7 tonnes in an un-manured field.[29] They also tried intercropping the nitrogen fixing beans with the wheat, and found that this benefitted both crops. We don't know if people in the Iron Age were aware of the benefits of this kind of farming, but we do know that there are signs of nitrogen depletion in many Iron Age soils – but also signs of manuring with nitrogen-rich animal dung. We can tell this by investigating the chemical makeup of buried soils and by the weed seed remains, and also by features we can see under the microscope.

One of the first experimental farms was not actually created for archaeologists, but for agronomists. Rothamsted experimental farm was established in 1843, to investigate the effects of the new inorganic fertilisers and to compare them to traditional techniques. These experiments are still running, and have an important contribution to make in the area of sustainable agriculture. Since the 19th century the research has expanded to cover a vast range of agricultural issues regarding pests, fertilisers, herbicides and plant genetics, to name a few. Experimental organic farms have also been established, for instance the Haughley Experiment, begun in 1939, which first demonstrated that there is seasonal variation in the nutrient availability in a soil. Experimental farms are central to our understanding about sustainability in past and present agriculture.

Economy and environment

One of the things that archaeology and anthropology have shown us is that there are many different ways of living off the same environment. Different cultures use different resources, and can live side by side on the same patch of land but in essentially different ecological niches. Hunter-gatherers before the Neolithic lived entirely on wild resources; with the onset of agriculture, a largely different set of resources were used, with farmers relying predominantly on domesticated crops and animals.[30] Pastoralists have traditionally lived near and traded with farmers, using the same ecosystem but in a very different way.[31] What these examples show us is that the environment we live in does not determine how we use that environment: our culture plays a large role in dictating how we live and which resources we use. Whether we are most influenced by our environment or our culture depends in part on where we live; in regions where there are many different resources available, different ways of life proliferate. By contrast, in extreme environments such as the Arctic, there is a limited range of resources, and people who don't use the full range of resources are unlikely to survive. One classic example is the case of medieval Greenland, where the Thule Innuit lived off of fish and sea mammals and wore warm furs, whereas the Norse settlers, who were their contemporaries, died out, probably due to their attempts to live like Europeans. The Norse wore wool instead of furs, and tried to maintain their cold and miserable domestic livestock in an arctic climate for which they were wholly unsuited. So it wasn't that Greenland couldn't support life, it was just that the Norse did not adapt to it very well.

Environmental Determinism is the idea that our environment dictates how we live. In some respects, our behaviour is moulded by the limits of possibility, but humans have an extraordinarily wide range of behaviours. It is this ability to make cultural adaptations that has made us such a successful species, and in the following chapters I want to discuss some of the behaviours, outlooks and technologies that could be adopted or adapted. Moving beyond the Green Revolution, the New Green Revolution is taking traditional knowledge and developing it in ways that are appropriate for the culture and environment. One of the most valuable lessons we learned from the Green Revolution – following the lesson that technology has to be appropriate – is the realisation that we cannot march into a country and tell the people what to do – even if they ask for our help. In fact, we can't do this in our own country either; changes introduced into *any* community are only going to work if you involve that community, asking them what would best suit their needs and coming up with solutions *together*.

The urgent need for more sustainable agriculture and architecture cannot be overstated, because the global population is growing and the climate is changing rapidly. There are natural variations in climate – something that climate change sceptics are keen to emphasise – but the changes that we are seeing today can ONLY be explained by the increase in greenhouse gases in the atmosphere. This is worth discussing in a bit more detail.

Climate change

When you do a Google search on climate change, you are initially faced with several pages of search results by climate change deniers. It is baffling that this should be so, but it is the case that there is a group of people to whom conspiracy theories are more interesting than evidence, and (perhaps less baffling) there are groups that are working to spread disinformation. While this may sound like a wild conspiracy theory in itself, we know that the tobacco industry ran disinformation campaigns for many years. Modern disinformation campaigners have become more sophisticated, and in order to understand the biases that are being presented to us, we have to read more widely in order to acquaint ourselves with the full range of evidence and opinion. When a topic falls outside our field of specialisation this can be difficult, but the way to go about it is to try to find reliable sources: peer-reviewed scientific journals, for instance, such as *Nature* and *Science*. It is also a good idea to read newspapers that represent both left- and right-wing opinion, because there are no 'good guys' anymore, assuming that there ever were. Newspapers leave out critical information that enables them to give a distinctive slant to what you are reading, without actually telling any falsehoods. The only way to get a clear picture is to read widely.

Regarding climate change: no climate scientists are querying the greenhouse effect. Any disputes in this area are over the importance of the individual gases that cause the heating. Nor is anyone disputing the fact that the climate has changed over geological time. Until recently, some scientists were still debating whether the earth is actually warming, but I don't believe any serious scientists are still disputing the fact that in the last hundred years we have experienced a sharp rise in global temperatures. To comprehend global warming, the first thing to understand is the greenhouse effect.

A greenhouse allows the sun's energy to penetrate as short solar rays, which are transformed into longer waves of heat energy, which are reflected back into the greenhouse by the trapping effect of the glass. However, when we talk about the greenhouse effect in terms of global warming, we are talking about a 'glass' made of gases, and this is where it gets a bit more complicated. The layer of gases actually absorbs heat from the sun, and although it reflects some of the sun's heat back out into space, it radiates most of it back down towards the earth. The increase in greenhouse gases means that not only is the troposphere – the layer of atmosphere closest to the earth – experiencing substantial warming, but we are also seeing cooling in the layer beyond (the stratosphere). This is because the greenhouse gases are trapping the heat in the troposphere, close to the earth.

The most significant greenhouse gases are, in this order, water vapour, carbon dioxide (CO_2) and methane; there are others as well, *e.g.* nitrous oxide and ozone, but they only contribute in a very minor way to the greenhouse effect. Some aerosols – notably black, carbonaceous aerosols– also increase global warming, while others reflect the heat back; the emissions from volcanic eruptions, for instance, are sulphuric, and absorb or reflect the sun's heat back out into space, and so contribute to global cooling rather than global warming.[32] Climatologists have to factor all of these effects into their models.

Carbon dioxide (CO_2) levels have a direct correlation with temperature change. Higher levels of carbon dioxide cause higher temperatures on the earth's surface. There are other factors that also affect global temperatures – obviously, there have been glacial periods, when vast sheets of ice covered nearly a third of the world's land mass, and there have been interglacials, which are warm periods between the glacials. However, the changes related to glaciations can be explained by changes in the amount of solar radiation received by the planet. The earth's orbit changes from elliptical rings to circular ones, and the earth's axis tilts back and forth sometimes, and it 'wobbles' as it rotates. These changes explain the glacial/interglacial cycle, but they do not explain the sudden, rapid rise in temperatures that we are seeing today. Only the measurable increase in greenhouse gases can explain today's climate change – and the increase in carbon dioxide can be directly linked to the Industrial Revolution.

Before the Industrial Revolution, levels of atmospheric carbon dioxide were at 280 parts per million (ppm); that is a baseline from which we can measure today's atmospheric carbon load. We have been burning coal for hundreds of years, but coal mining and use increased exponentially from the mid-19th century, when coal was burned not just for household heat and cooking but for industry and transportation. In 2014, carbon dioxide levels in the atmosphere reached 400 ppm, which is unprecedented in recent geological history; by 2015, 400 ppm became the global average.[33]

The Intergovernmental Panel on Climate Change (IPCC), which is made up of over 1000 of the leading climate scientists, is aiming – desperately – to prevent CO_2 levels from exceeding 450 ppm, because at that point, the 'centre ground' climate models (*i.e.* the fairly cautious but not extremely cautious models) predict that global temperatures will increase by another degree, which means that the earth will have experienced a 2°C temperature rise since the mid-19th century. If you wish to understand what that means for the planet, I recommend Mark Lynas's book, *Six Degrees*, which uses our understanding of global temperatures and ecology to outline how each degree in temperature rise will affect the world's ecosystems. Read it and weep.

We have records of the temperature from the mid-19th century, and we can see that from the 1910s to the 1940s the average global temperature rose by 0.35°C. From the 1970s to the early 21st century the temperature rose by around 0.55°C.[34] All together, we have seen a global rise of 1°C since the beginning of the 20th century, but while temperatures were rising by 0.1°C per decade through most of the 20th century, more recently we've been seeing a rise of 0.15°C per decade. So, how do we know about temperatures in the past, before we have thermometer records?

How do we know about temperatures in the past?

We have a number of ways of assessing past temperatures, called 'proxies', which are indirect measures. Tree rings, for example, provide precise dating evidence through dendrochronology, but they also provide an indication of temperature, based on the size of the annual growth rings. Coral reefs and ice cores provide a record of the ratio of oxygen isotopes, which are a particularly important indicator of past climate. There are three oxygen isotopes, two of which are used in climate science: ^{16}O and the ever-so-slightly heavier ^{18}O. During periods of warmer climate, the lighter form of oxygen is evaporated more readily, leaving the ocean and other bodies of water enriched in the heavier ^{18}O. Oxygen isotopes are absorbed into the calcium carbonate shells of microscopic sea creatures, which die and deposit their shells on the ocean floor, and so provide a stratified record of past sea temperatures. Oxygen is also trapped in ice near the poles, and is also conserved in the calcium carbonate of speleothems – the mineral deposits that form in caves, including stalactites and stalagmites.

Lake sediments also provide clues to past environments and climate, and we also have historical records for the more recent past. Data from ice cores is particularly important, because ancient air is trapped in bubbles in the ice caps of the Arctic and Antarctic, so the greenhouse gases from the past can be measured directly. This research has shown that the levels of carbon dioxide, methane and nitrous oxide in the atmosphere all show an extraordinarily sudden rise in the mid-19th century.

In 1998, a long-term climate model was created, which included not only recent temperature records, but also the information from climate proxy data, taking it back to AD 1400.[35] The model was refined in 1999, to take it back to AD 1000 (Figure 1.5).[36] The graph is the result of this important work; it is called the 'Hockey Stick Graph', because of the shape (the 21st century shoots off at an angle from the 'handle'). One of the important aspects of the graph is that it includes error bars (the grey shading), which show the range of possible error; in this instance, the data has a 95% probability of falling within this margin of error. You can see that there is a large margin of error before about 1600 – but it is also clear that what is happening now is like nothing we have seen over the past 1000 years, even if we are wildly over- or under-estimating the temperatures in the past. Even the Medieval Warm Period, when temperatures were slightly warmer, did not show such a dramatic change as the one we are seeing today.

Climate models have often proved to be too cautious and conservative in their predictions, largely because of the uncertainty of the level of 'positive feedback'. This is an ecological term for what happens when one factor causes another, which causes another, and so on – in a sort of domino or spiral effect. More CO_2 in the atmosphere

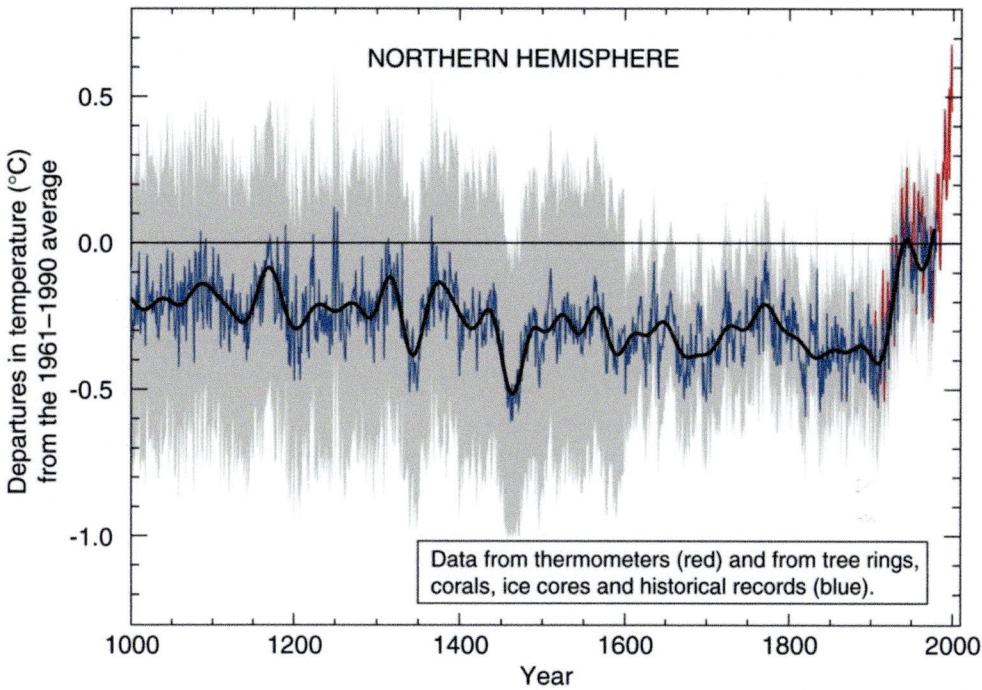

Figure 1.5: The Hockey stick graph. (IPCC 2001)

causes warming of the earth, which causes more water to evaporate. Water vapour is a greenhouse gas, so more water vapour causes hotter temperatures... and so the problem gets progressively worse.

What will this mean for the planet? The Greenland ice sheet is melting and sea ice thickness and extent in the Arctic have both decreased. This is causing the sea level to rise, and it will rise by at least half a metre by 2100 – but it could rise by up to 10 m if the Greenland and West Antarctic ice sheets melt.[37] We don't know how bad it's going to be, because glaciers don't melt in a steady, linear fashion.

Could all this just be some kind of mistake?

Sadly not.[38] There is no model that explains away the increase in carbon dioxide in the atmosphere over the last few hundred years, and the resulting increase in temperature. There is no 'natural' explanation – none whatsoever – and we need to concede that climate change is real, and that it is a man-made problem.

The future

I want to stress that traditional knowledge is not a monolithic block of unchanging information; it grows and develops as each generation performs new experiments, and it develops as new ideas are traded with other cultures. At the moment there are extraordinary new developments in technology, medicine and communications, and many of these can be combined with simple technologies and traditional knowledge to create cheaper, more sustainable solutions to our current problems, not just in developing countries but also in the West. We are already producing enough food to feed the world[39], and there is no reason to believe we cannot find better ways to grow, harvest and distribute all this food.

It is also important to recognise that few people wish to live in a museum, and some forms of modern technology have got to be incorporated into traditional societies – not least, advances in modern medicine. As more and more people around the world are exposed to the Western way of life, many will – quite reasonably – wish for the benefits of electricity, modern medicine and electronic devices and communications, and will also wish to adopt the higher levels of meat consumption that almost invariably increases with growing wealth.

I will leave the discussion of sociological and economic issues to experts in these fields; in this book, I want to focus on practical solutions to environmental management issues, and I particularly want to discuss ways of working *with* nature instead of against it. I want to consider a number of local and regional solutions that ensure that there are not winners and losers, but rather, developments in which everybody wins – people, the economy, the local environment and the planet.

Notes

1   *e.g.* Diamond 2005.
2   *e.g.* King 1911.
3   Guttmann-Bond 2010.
4   Carson 1962.
5   Schumacher 1973, 149.
6   Munn 1992.
7   Pretty and Bharucha 2014.
8   Scialabba and Hattam (eds) 2002, 127.
9   Unnayan Bikalper Nitinirdharoni Gobeshona (the Policy Research for Development Alternatives).
10 Scialabba and Hattam 2002, Chapter 4 box 8.
11 Scialabba and Hattam 2002, 155.
12 Scialabba and Hattam 2002, 155.
13 Pretty and Bharucha 2014.
14 Thought for Food. *New Scientist* 174 (2343), 18 May 2002, 34; MacKerron *et al.* 1999.
15 Conway 2005.
16 Pretty 1998a, 84.
17 Hillman 1981; Reynolds 1995.
18 Derpsch 2004.
19 Montgomery 2007, 211.
20 Derpsch 2004.
21 Kassam *et al.* 2015.
22 Jules Pretty, pers. comm.
23 Pretty and Barucha 2015.
24 Gliessman 2007, 11.

25 Pretty and Barucha 2015.

26 Reimer *et al.* 2013.

27 Serpell 1989.

28 Reynolds 1995.

29 Reynolds 1995.

30 Richards *et al.* 2003.

31 Some hunter-gatherers today are also able to trade with local farmers, but this opportunity was not available before the Neolithic, when there were no farmers.

32 Mann and Kump 2015, 14–19.

33 World Meteorological Organization: Globally Averaged CO2 Levels Reach 400 parts per million in 2015. Available online at: https://public.wmo.int/en/media/press-release/globally-averaged-co2-levels-reach-400-parts-million-2015 (accessed 22 June 2017).

34 IPCC Working Group: Can the warming of the 20th Century be explained by natural variability? Available online at: https://wg1.ipcc.ch/publications/wg1-ar4/faq/wg1_faq-9.2.html (accessed 22 June 2017).

35 Mann, Bradley and Hughes 1998.

36 Mann, Bradley and Hughes 1999.

37 Mann and Kump 2015.

38 Mann and Kump 2015; IPCC Working Group: Can the warming of the 20th century be explained by natural variability? Available online at: https://wg1.ipcc.ch/publications/wg1-ar4/faq/wg1_faq-9.2.html (Accessed 22 July 2017).

39 FAO 2017a, 4.

2

Wetlands and wetland agriculture

Wetlands in the modern world are generally regarded as an obstacle to commercial agriculture, apart from in the rice-growing regions of Asia. As a consequence, 50% of the world's wetlands have been drained since the late 19th century.[1] In Europe, around 66% of our wetlands have been lost in the last hundred years or so, most of them having been drained for agriculture. Drainage projects range from the small scale improvement of boggy ground to the reclamation of vast areas like the Dutch polders and the Cambridgeshire fens.

It has not always been this way. In pre-industrial Europe, low lying and seasonally flooded hay meadows were a valuable resource, because hay was the essential fodder crop that allowed farmers to overwinter domestic livestock, including the horses and oxen that pulled the ploughs. After farming was mechanised, these species-rich and biologically diverse hay meadows were no longer a necessary part of the agricultural landscape, and wet landscapes came to be regarded as waste ground. Mechanisation has removed the need for traction animals in the developed world, and has made hay meadows largely redundant, but today we are increasingly recognising that wetlands and meadows are important because of their biodiversity, and floodplains are particularly important because of their role in flood prevention. For these reasons, wetlands, meadows and floodplains are now being reinstated in many areas.

Looking back in time, but also looking at the wider, global picture, it becomes evident that there are actually many ways of living in and on and from the wetlands, many of which don't involve draining them and ploughing them flat. In much of south-eastern Asia, wetlands are so essential for growing rice that there are rice paddies crowded into every available open space. On the train from Tokyo to Kyoto, for instance, I was astonished to see tiny rice paddies in the gaps between the buildings in urban areas, which in European cities would simply have been left as vacant lots.

Before farming was introduced, wetlands provided our hunter-gatherer ancestors with waterfowl, small mammals and fish; the earliest evidence for fishing goes back to at least 90,000 years ago in Africa.[2] Archaeological evidence from the coasts and wetlands of Northwest Europe includes fish traps, fish hooks, harpoons and fishing nets dating back to the Mesolithic period, about 8000 years ago.[3] The wetlands of northern Europe were well populated in prehistory, with fish being an important source of food. In North America, fish hooks from around 9500 years ago have been found,[4] and wetland-based cultures can still be found around the world today, but many (such as the Marsh Arabs in Iraq) have had long, usually unsuccessful struggles with dryland farmers, industry and unsympathetic governments.

In Europe, prehistoric wetland communities farmed the dry land surrounding their lake villages and island settlements,[5] but it is also possible to make a good living by actually farming the wetlands themselves, using only simple technology and local materials to transform lakes and swamps into uniquely productive land. In Central America, people have been farming the swamps using the same, sustainable methods for hundreds, even thousands of years. In South America, the ancient methods had been forgotten by the local people, and have only recently been rediscovered by geographers, agronomists and archaeologists.[6]

The ancient wetland farming systems are now being reintroduced, and are turning out to be more productive per hectare than the most modern farming techniques in the region. Lands that were mapped as barren wasteland or rough grazing have been made intensively productive by simple techniques requiring no modern technology, no tractors, no chemical fertilisers and no chemical pesticides. I will begin with a discussion of a system that has survived from antiquity into the present: the famous 'floating gardens of Mexico'.

The floating gardens of Mexico

Tenochtitlan was the capital of the Aztec Empire, and at its peak period (around AD 1520) it is estimated to have had a population of 150,000 to 200,000.[7] The city was located in the Basin of Mexico, a large valley in the high mountains at the centre of Mexico. The 'Basin' is actually 2240 m above sea level (about 1 ½ miles), and is surrounded by towering volcanoes reaching up to 5500 m. Before the Spanish arrived it was filled in the rainy summer months by a vast, shallow lake which the Aztecs called the Lake of the Moon.[8] The lake was only about 1–3 m deep, and in the drier winter months the water table dropped and it divided into 6 separate lakes.

The northernmost lakes were Zumpango, Xaltocan and (under Spanish rule) San Cristobal, and the large central lake was called Texcoco. The northern and central lakes were fed by rivers which flowed over steeply sloping volcanic rocks with a high sodium feldspar content.[9] The rivers carried this mineral into the lakes, where it dissolved into salt. The mountains to the south, by contrast, were covered in permeable lava, so the water seeped down into the ground and emerged as pure, sweet springs which filled the southern lakes, Xochimilco and Chalco.[10] Tenochtitlan was founded in Lake Texcoco, which had a mixed source of water.

The cities of Xochimilco and Chalco were established by the Xochimilco and Chalco Indians in the 12th century,[11] and these early farmers had a number of environmental difficulties to cope with in this mountainous region. The sheer height of the valley above sea level posed a number of problems, particularly the problem of a cold, short growing season, and added to this were the variable rainfall, hard frosts, relatively infertile soils, and – in the northern and central lakes – saline water. Taken altogether, these are some formidable obstacles to successful farming. Even so, in the centuries before the Aztecs arrived, the early farmers got around these problems by using an inspired method of land creation: fences of reeds were sunk into the lake bottom and were filled with layers of clay, silage, aquatic plants and manures to create pockets of land in the shallow lakes.[12] Willow trees were planted along the edges of the new, artificial islands, partly for shade and for firewood, but also serving to control erosion.[13] This type of man-made island is called a 'chinampa', from the Nauhatl word for reed basket[14] or for an enclosure of reeds.[15] Chinampas date back to as early as AD 800,[16] and had become an important feature by AD 900,[17] but were probably not extensive before the Early Aztec period, around AD 1300.[18] Their use appears to have increased along with the rising Aztec population.

Historical accounts say that the chinampas were built on floating mats of vegetation which were made into floating islands by piling on layers of mud from the lake bottom.[19] Joséde Acosta wrote in the 16th century that the chinampas actually floated and could be towed about by the farmers.[20] There are no known floating chinampas today, and the idea that there ever were has been fairly soundly refuted by Gene Wilken, who has rigorously studied the area, but the idea of floating islands is so appealing that it still hasn't quite died out.[21]

Chinampas built in historical times are usually 2–4 m wide and 20–40 m long, with ditches on at least three sides.[22] The soils are fertilised with mud and aquatic vegetation from the adjacent ditch or canal bottoms. This material is very rich in nitrogen, the key nutrient required by maize crops. Household waste is also added, and in pre-Hispanic time human excrement was also used,[23] a practice that is perfectly safe so long as it is composted properly to kill off any harmful bacteria such as e-coli. Bat guano was also brought in by slaves from the caves of Morelos.[24]

The chinampa soils are fine enough for capillary action to bring moisture up to the plants, so the soils are well watered, and the proximity to the water table extends the growing season in the spring and autumn by inhibiting the frost. The soils cannot be sustained indefinitely, however. Salts from the lake water and mud gradually accumulate in the chinampa soils, and eventually they become too saline. Also, the repeated additions of lake mud eventually raise the crops too high above the water table, and they can no longer get enough moisture. These problems are addressed by periodically removing the soil and replacing it with fresh mud from the lake bottom. Crop rotation is also practiced, to maintain fertility. The chinampas are extremely productive, even occasionally producing more per hectare than the local agricultural research stations.[25] They can generate 3–5 metric tonnes of food per hectare, and produce two crops per year. This is an extremely productive system, even by modern standards.

In the 1440s the city of Tenochtitlan was subject to a number of severe floods, and the engineer King Nezahualcoyotl, together with King Montezuma I, built a dam across Lake Texcoco, which stemmed the periodic floods.[26] This served both to protect the chinampas and to reduce the salinity of the lake, which meant that the lake water could be used more intensively for irrigation, and the soil would not have to be replaced as frequently. The king also built drawbridges and sluice gates to regulate water and transport.

One of the most important engineering works of this time was the construction of an aqueduct, built in 1440–1468.[27] This brought fresh spring water along a causeway leading out over the lake to the city. A second aqueduct was built later by the Emperor Ahuítzotl, but initially this proved to be too successful, bringing so much water that the city was subjected to sudden, violent floods. The problem was addressed by the sacrifice of some high officials, whose hearts were cut out and thrown into the canal, together with some valuable artefacts. Interestingly, the flow of the spring then diminished.[28]

Apart from this questionable understanding of cause and effect, the Aztecs worked well with the resources available to them. The ecology of the complex chinampa system was maintained for centuries without any diminution in the soil quality, because the soil could simply be replaced when it became too saline. This balance was destroyed when in 1521 the city of Tenochtitlan was captured by the Spanish Conquistador Hernán Cortés, after a 79 day siege. Cortés founded the modern Mexico City on the ruins of Tenochtitlan, breaking up the Aztec dykes and using the stone to build the new city. This resulted in huge floods, which were not abated by the Spaniards' misguided attempts to drain the lake.[29]

The problem was compounded by the imported Spanish sheep that were set to graze on the hill slopes. Grazing animals can do a lot of damage to the soil, because their hooves break up the vegetation mat that keeps the soil in place, and it's a particular problem on sloping land. The result was that great influxes of sediment were carried down the mountainsides and into the lakes whenever there were heavy rains. The sediment raised the level of the lake beds, which raised the level of the floods. After a particularly bad series of floods in 1604 and 1607, the Spanish dug a canal to drain the lake. Sixty thousand Indians were put to work on the project, during which an estimated 10,000 to 12,000 died of illnesses and a further 10,000 died in accidents – an unenviable Health and Safety record.[30]

The Spanish had hoped to farm the lake bed once the lake was drained, but their efforts were confounded by the discovery that it was too saline. Nothing would grow on it, not even rough pasture grass, and furthermore the problem with flooding continued, even after three rivers were diverted.[31] Large areas of chinampas were lost, and now the Spanish had problems with dust storms from the unvegetated wasteland.

In the early 20th century the springs that feed the lake began to be tapped to supply clean drinking water to the city, and the lake levels dropped once more (resulting in the ongoing problem of subsidence of the city's buildings). When the farmers (called Chinamperos) protested about this, the government 'assisted' them by redirecting partially treated sewage water mixed with industrial waste into the canal system.[32] Many plant species were unable to survive in the polluted water, and species diversity plummeted. The water table is also sinking because of deep wells that have been dug for drinking water, and another problem has been the attempts to consolidate chinampas and to farm them with tractors. This has reduced soil quality, presumably because the tractors compact the light structured soils, reducing the capillary action that made them so ideal. In 1973 the canal water was deemed unsuitable for irrigation, and by then all the fish and amphibians that used to populate the canals had died out.[33]

In the 1980s a coalition of chinampa farmers, environmental groups and ecologists from the local University got together to protest against the loss of the chinampas. In 1993 the Xochimilco Ecological Park was set up to save the chinampas and to restore the ecology that made them so successful, and the floating gardens of Xochimilco were proclaimed a World Heritage Site in 1998. It has been estimated that the remaining chinampa fields could supply Mexico City with one quarter of the vegetables required – and such a scheme would also save a great many food miles.[34] The city also benefits from the tourism that these very beautiful (although non-floating) gardens attract (Figure 2.1). In spite of these benefits, only 160 ha of the surviving 3000 ha

Figure 2.1: The chinampas today. (Photo © Journey Mexico)

of chinampas are actually protected, and the water quality is still poor. A report in the late 1990s noted that 9000 new houses a year were being built in the supposedly protected canal banks, and raw sewage was still occasionally pumped into the canals.[35]

Amazonia

The South American rainforests are rich and fertile ecosystems, but when they are cut down on a large scale they cannot regenerate. This is why conservationists are so appalled by the destruction of the rainforests – if the forests are cleared, they are naturally replaced with savannah, a harsh, arid landscape of scrub and rough grasses. The reason for this is the nature of the soils. Tropical rainforests produce a huge amount of organic material, which all returns to the soil when plants and trees die and decompose. If the trees and vegetation are cleared, there isn't enough organic matter going back into the soil, and it loses its nutrients and its nice, spongy structure very quickly.

Because of the inherent infertility of the rainforest soil, it is regarded as terrible farmland, and consequently the regions with such soils have very low populations. Modern rainforest tribes who practice farming generally use slash and burn agriculture, which involves cutting and burning a clearing in the woods. The ash from the burned vegetation is very high in nutrients and is also very calcareous, so it combats harmful soil acidity. This produces a great surge in the soil's fertility, but unfortunately it declines again very quickly. The clearing can only be sown for a few years before the soil becomes exhausted and has to be left to regenerate. This kind of farming may sound destructive, but it's actually quite sustainable so long as the population is low and not too much land is cleared. The problem is when too many clearings are made and they aren't given long enough to regenerate, and then the land deteriorates. That is what is happening now.

The natural infertility of the soils in this region poses a problem to archaeologists and historians. When the Spanish arrived in Amazonia, they reported finding a large population governed by complex chiefdoms – but how could such an infertile soil have supported such a large population? Some archaeologists have argued that the environment was simply too harsh, and the Spanish reports must be mistaken.[36] This argument is based on Environmental Determinism, which is the idea that the nature of an environment and the type of resources it produces determine how many people it can support (and to some degree the type of culture, although this idea has fallen out of favour). The idea has some merit, in that there are natural limitations to vital resources in extreme conditions, but often it does not take into consideration all the different ways that people have of adapting to an environment.

The environmental determinism argument has been adopted in the debate over past populations in the Amazonian rainforest. Meggers argued in 1954 that 'the level to which a culture can develop is dependent upon the agricultural potentiality of the environment it occupies'[37] – but anthropologists and archaeologists in the 1950s didn't fully understand the degree to which people can improve their agricultural potential using simple, pre-industrial methods. Meggers argued in 1971 that the soils of Amazonia are inherently infertile and can't support farming, and therefore there can't have been very many people living there and there certainly can't have been cities such as the Spanish Conquistadors described when they conquered South America.[38] An interesting debate began when new evidence emerged from an unexpected source – evidence that most archaeologists now believe demonstrates that a large population of sophisticated farmers once lived in the South American wetlands.[39] Looking back, this is how the debate began:

In the 1960s, a group of oil exploration geologists discovered a strange land form in the Amazonian swamps. There appeared to be thousands of hectares of artificial islands, for which there was no obvious natural explanation – but if they were man-made, it wasn't at all clear what they were for. A number of aerial surveys were undertaken by interested colleagues and by a young PhD student, William Denevan, who decided to write his thesis on the cultural-historical geography in this region. As he flew over the area for the first time, he realised that what he was looking at were the ruins of an ancient civilisation, in an area where he had expected only swamps. This led to four decades of research by Denevan and a number of others, undertaking aerial surveys to map the extent of the islands and carrying out excavations to try to work out how old the islands were and what they were used for.

Their research was in an area called the Llanos de Mojos, a huge savannah in the Amazon Basin in northern Bolivia. Most of this area is seasonally flooded grassland, which is wetland for 4–6 months of the year. There are also areas of scrub, forests and islands. The Spanish arrived in the 16th century, bringing with them European diseases which wiped out a large proportion of the people.[40] Slave raids then prevented any sort of stability returning to the region. The land was settled by Jesuit missionaries in 1682, and in 1767 the Jesuits left and the cattle ranchers moved in. The ranchers remain to this day, although the cattle numbers are declining, and the only agriculture that takes place is of the slash and burn variety, in the woodlands. So who built the raised fields?

The archaeologists, geologists, geographers and engineers proceeded to map and classify the fields from the air. Their conservative estimate was that the fields covered 6000 ha in the western Beni region of Bolivia, but they believe the actual figure is at least double.[41] Many were badly eroded and were not obvious from the air, and many more had become swallowed up in the forest which grew up after the population declined. No one now remembers the fields ever having been cultivated, but Denevan believes they could easily have supported the large prehistoric population recorded by the Spanish.[42]

The most substantial type of field were the platform fields, which were 5–20 m wide and up to 335 m long. They were spaced 3 m to 100 m apart or more, and were built in clusters of several hundred. The clusters of ridges are at odd angles to one another, which suggests they were built up piecemeal as the population grew, and not as part of a massive planned landscape. Ridged fields were much narrower, measuring 1.5–6 m wide and 6–300 m long. Mound fields were made up of regularly spaced mounds measuring *c.* 1.5 m in diameter, built in long lines. This type of plot occurred in more marginal areas out on the open pampas and away from other field types, possibly because this land was too deeply flooded to build the large ridges seen in other areas. The fourth type, gridiron fields, were rectangles enclosed by ditches. As well as ditches there were causeways, possibly created to control the flow of water to different parts of the field systems, and there were canals throughout the system that could also have been used for transport.

Excavations by John Walker, an American PhD student, were undertaken in the northern part of the Mojos, around the Rio Iruyañez. Radiocarbon dates from his excavation of the settlements provided dates of AD 410–620 and AD 1275–1645.[43] In 1990 further work was carried out by Clark Erickson, who came up with radiocarbon dates of AD 1 and AD 1200.[44] Why such a long time span? Radiocarbon dates are taken from organic material such as bone or charcoal, and they have to be interpreted based on where on a site these datable materials are found. Because farming settlements tend to be inhabited for long periods of time, you can get a broad range of dates from the same ploughsoil, because people kept adding fresh fertilising material and ploughing it in, so the same soil was turned over again and again. You can also get a range of dates from the different parts of the site, with earlier layers of soil and old backfilled rubbish pits being buried by later deposits. In this instance, the dates from the two excavations at the Mojos indicate that the settlements associated with the fields are very early, predating Columbus by over a thousand years, but with a final date occurring during the colonial period.

The locals say that the pampas (grassland) soils are too heavy, waterlogged and infertile to cultivate, and it's difficult to keep the savannah grasses from encroaching on the farmland. Modern analysis of the soils in the ancient raised fields confirm that they are a grim prospect for agriculture, characterised by leaching (loss of minerals), mottling (an indication that they are seasonally waterlogged, *i.e.* they get no oxygen for part of the year) and hardpan formation (this is when soil minerals wash down and accumulate in a solid crust some way beneath the topsoil).[45] However, if the soils were fertilised and maintained in prehistory in the way that modern farmers fertilise the chinampas, then the soils here would have been excellent. They would have been well watered, well aerated and well fertilised, and also protected from the frosts by the proximity to the water table.

To test this hypothesis, Erickson and his students set up an experiment to test the productivity of the fields. He had some outstanding results, growing 25 metric tonnes of manioc per ha and 2 metric tonnes of maize per ha, even including the standing water of the ditches in his calculation of land area.[46] The locals were astounded – but as most of the savannah is controlled by ranchers, there wasn't much interest in reinstating the fields. Also, the land has remained depopulated since the arrival of the Spanish, without even a road to link it to the outside world until the 1980s. For archaeologists this has been a blessing, because the relative isolation of the region meant that the fields hadn't been ploughed flat, as they have been in more populated areas. The preservation has enabled archaeologists to gain a good understanding of the system, and though it is unlikely to be put back to work in this depopulated area, still it can be used elsewhere in South America.

The Andean highlands

High in the Andes, between Peru and Bolivia, there is a large area of wetlands where at least 82,000 ha of prehistoric raised fields survive around Lake Titicaca[47] (Figures 2.2 and 2.3). The fields were created in the same way as those in the low-lying Amazon basin: parallel canals were dug and the earth was piled up in the middle to make raised beds for cultivation. The Lake Titicaca fields are generally 4–10 m wide and 10–100 m long, but unlike the fields in Amazonia they can still be seen on the ground, and not just from the air. As in other regions, the raised fields benefit from the proximity to the water, which creates a warm microclimate and prevents frost damage, and the rich lake muds make superb fertiliser. Nevertheless, the soils are classified as 'marginal' by the regional soil surveys, and the fields are unfortunately being ploughed flat for wheat cultivation.[48]

Excavations by Clark Erickson and his team have provided dating evidence both for the fields and for the settlements scattered among them. Pottery in the fields dating to 1000 BC has been recovered, and dates from the settlements indicate that they lasted from 1000 BC to AD 300, although there could have been breaks in the occupation. Settlement in the area was then re-established between AD 1000 and 1450.[49] The settlements are on mounds which built up over many years of building and rebuilding in mud brick – just like the world's first cities, the 'tell' or mound sites in the Middle East, which formed in much the same way. To this day the mounds are regarded as ideal places to live, because they are raised up above the level of the floods.

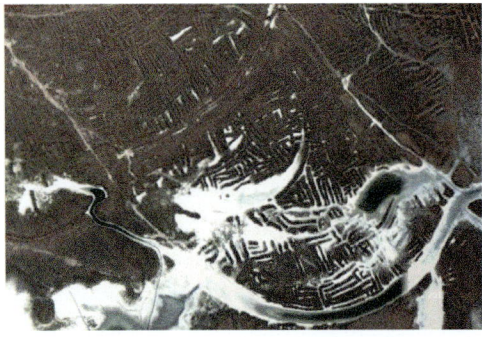

Figure 2.2: Satellite image of raised fields on the edge of Lake Titicaca. (Photo © Google Earth)

Figure 2.3: Satellite image of a drier area of the raised fields on Lake Titicaca. (Photo © Google Earth)

The prehistoric people around Lake Titicaca kept llamas and alpacas, and the wetlands provided fish and wildfowl as well as agricultural land. The bones of guinea pigs were also found in the excavations. Evidence from pollen preserved in the lake and wetlands indicates that potatoes and quinoa (a South American grain) were grown here, and in fact this may be the first place in the world that potatoes were grown.[50] Altogether, the prehistoric economy looks very much like the economy of the Indians living there today, but with one key exception: the prehistoric raised beds were much better farmland than the sterile hill slopes that are farmed today.

The site that Erickson excavated went out of use before the Spanish arrived, but there was no environmental reason for this to have happened, and he argues that there must have been a social or economic reason.[51] Most of the population seems to have shifted to a different area, perhaps a new ceremonial centre, but the fields may not have been abandoned altogether. Since there didn't appear to be any environmental reason for the fields' abandonment, there seemed to be no reason why they shouldn't be reinstated. Excavations provided information on the original canal depth and ridge spacing, and it was a straightforward task to rebuild an area with the help of local Quechua farmers using their traditional tools, the Andean foot plough, hoe and clod breaker. These simple tools proved to be ideally suited to the task. The modern tools are made of metal rather than stone, but are otherwise the same design as the prehistoric ones; we know this partly because thousands of basalt hoe fragments have been found in the ancient raised fields. To some degree the continuity of design may be simply because it's the most practical. The Andean foot plough, which is used to cut turves, is almost identical to the delling spade that was used to cut turves of peat in Scotland in recent historical times. A similar tool is still used in Ireland for peat cutting today.

The experimental raised fields were cultivated using only the prehistoric methods: they were fertilised with lake mud and plants from the canals, including nitrogen fixing blue-green algae (which provided nitrogen). Fish living in the canal also add nitrogen and phosphorus in their excrement. This becomes incorporated into the canal mud, which is dredged up and spread onto the fields.

The resulting crop took everyone by surprise, and for two reasons. Firstly, the yields from the experimental farm were so high and the vegetables were so large that the local farmers were convinced that the archaeologists had used some amazing new fertiliser, which they hadn't. The average potato yield from the plots was 10 metric tonnes per hectare, compared with the regional average of 1–4 metric tonnes per ha.[52] Also, the raised beds can be cropped continuously, without exhausting the soil, and the lake mud costs nothing.

Secondly, in 1982, during the course of the experiment, there was a severe frost that badly damaged the crops in the modern fields around Lake Titicaca – but the crops in the raised beds suffered only minimal damage, thanks to the temperature regulation provided by the canal system. The night time temperature over the raised fields was 2°C higher than the temperature over the modern dryland fields, and the temperature over the canals was even warmer because the water acts as a heat store. This not only protects the crops, but also extends the growing season.

The system of wetland agriculture was put back to work by the native Quechua and Aymara peoples in Bolivia and Peru, who by 1990 had reinstated between 500–1500 hectares of prehistoric raised fields.[53] They have not just benefited economically from the endeavour: they make a point of impressing on visitors the pride that they feel in this link with their ancestors, who created a viable, sophisticated and sustainable system in a difficult and marginal environment.

The development since then has proceeded sluggishly or not at all; individual farmers who own their own raised fields continue to farm them profitably, but the community efforts have broken down due to interpersonal tensions and issues of land ownership.[54] NGOs also instituted a 'top-down' approach, which provided financial incentives above and beyond the incentives of a more productive farm.[55] Such efforts are now renowned for their failure, as they tend to not take into consideration the farmers' needs or the social structure of the local people, and they make the people beneficiaries instead of participants, so they feel no ownership or control over the projects.[56] There has also been political unrest, and the NGOs who ran the projects made errors such as emphasising the appearance of the fields instead of the ecological balance. Nevertheless, Erickson's extensive research and successful experiments have demonstrated that the raised fields can support a large population using only local resources. The farmers relying on industrial methods have the capacity to grow far more than they are currently producing, and without the costly necessity of fallowing (*i.e.* leaving the land uncultivated, so that it can recover its fertility naturally) and with no costly inorganic inputs.

Water meadows in Britain

The South American systems relied on the proximity of water to extend the growing season, which occurs because standing water retains the heat from the day and releases it slowly overnight, keeping the crops from freezing. This happens naturally along rivers and floodplains, and the value of such meadow land was recognised by farmers from at least as early as the Roman period in Britain, but probably also in prehistory. In Oxfordshire, for instance, there are several low-lying prehistoric sites that seem to have a specialist economy based on horse breeding.[57] The Iron Age settlement at Mingies Ditch, on the floodplain of the Windrush Valley, had an unusually high proportion of horse bone, including the bones of young foals.[58] The low-lying Iron Age sites of Farmoor, Northmoor and Appleford showed a similar bone distribution pattern, and it is believed that these valley sites were breeding horses. Nearby sites on the uplands, by contrast – for example Danebury Hill Fort and Faringdon[59] – seem to have had horses imported, because there were adult horses on site but no evidence for young foals. It seems likely that specialist upland and lowland sites were trading in Britain in the Iron Age – and the regularly flooded meadows in the valley bottoms may have played an important role in this diversification and trade by providing rich meadows for pasture and hay for overwintering the stock.

Meadows are defined as 1) areas of grassland used for hay, and 2) low-lying areas near rivers.[60] They are distinct from pasture because the grass is harvested instead of being grazed; in this way fodder can be stored to feed domestic animals over the winter, when grass is scarce. We know that the Romans had hay meadows because of the discovery of Roman scythes, and we also have linguistic evidence in that the Romans had distinct words for pasture and for meadow.[61]

The Saxons also had distinct words for meadow and pasture, and the Old English word 'mede' (deriving from 'māwan', to mow) is the origin of the word meadow, and occurs in names like Runnymede. The word 'ley' is also thought to be derived from Old English and seems to be another word for meadow. The first written evidence for meadows is in AD 679, and by the time the Domesday book was written in 1086 there were over 10,000 entries for meadow, which then made up about 1.2% of the land area in England.[62] Meadows were important for supplying hay to overwinter the oxen that pulled the ploughs, and as arable crops became more important, meadows also became more important because of the necessity to feed the working animals. By 1250, England was probably about 4% meadow.[63] As sheep became more important in the later Middle Ages, meadows became even more important because they ensured the survival of large flocks over the winter.

By the Middle Ages, English farmers were using irrigation to create artificial meadows. These systems became more complex over time, eventually developing into the elaborate 'water meadows' that became an important feature of 17th–19th century agriculture. In the 14th and 15th centuries there are accounts that suggest that there was controlled flooding of meadows using banks, ditches and hatches that could be lifted or shut to control the flow of water[64] (Figures 2.4 and 2.5). By 1523 farmers had created an elaborate system of channels, called leats, all carefully levelled to guide the flow of water from the river and through a system of ridges and furrows.[65] The system was recorded in Rowland Vaughan's *The Most Approved and Long Experienced Water Workes*, published in 1610.

Water meadows were created by 'floating' the low-lying floodplain, a process that released just enough water to gently wash over the soil surface, preventing the land from freezing. By carefully flooding the meadows four to six times a day, the moving water – which ran through the grass but not over it – raised the temperatures above 5°C and so provided conditions for the grass to begin its spring growth.[66] The flow was controlled to ensure that the water did not stagnate, which would cause it to become oxygen deficient, which would kill the grass. In addition to preventing the land from freezing, the water that ran through the meadow added to the fertility of the soil, because it contained runoff from arable fields and was therefore high in nitrogen and phosphorus.

The value of these systems was that they could increase the hay crop by at least 4 times,[67] and like the chinampas they also lengthened the growing season. Traditionally, farmers' hay stores would be running low by April, but water meadows ensured an early spring growth of grass for livestock, so that they functioned as pasture as well as providing hay. The fields could be drained in late February, and livestock – usually sheep – could be put out onto these fields as early as March. When the upland pastures were ready, the sheep could be transferred and the meadows returned to flooding in order to produce the winter's hay crop. In the autumn, the sheep or cattle could be returned to the water meadows to graze the final crop before overwintering indoors.

By 1800, water meadows covered between 6000–8000 of hectares in Wiltshire alone, but most have since been destroyed by modern agriculture.[68] In the 1950s and 60s, the Ministry of Agriculture, Food and Fisheries actually paid farmers grants to flatten old water meadows, which were regarded as a burdensome anachronism. New grass varieties and chemical fertilisers were

Figure 2.4: Water meadows at Harnham, Wiltshire. (Photo courtesy of the © Harnham Water meadows Trust)

Figure 2.5: Water meadows from the ground. (Photo © Hadrian Cook)

used to improve the upland pastures, which were simpler and less labour-intensive to manage. Nevertheless, you can still make out the remains of the old ridges and furrows along certain river valleys in the south of England if you look carefully. Look for the channels that support the system, leading water into the furrows and eventually back out to the river.

Water meadows are now being restored, managed and conserved, but largely out of historical and environmental interest. In the UK, it is assumed that low-lying land is not compatible with agriculture – but the industrialised agricultural landscape in the Netherlands teaches us otherwise.

The Dutch polders

The lowlands of the Netherlands are formed by the river deltas that drain the European Continent, with much of the low-lying land being in fact the delta of the Rhine River. This particular geography has given this region its distinctive character and its distinctive engineering responses to flooding. The drained lands of the Netherlands were not created all at once, but piecemeal, beginning in the Middle Ages. Polders – the fields reclaimed from the sea – were created by building banks, or dykes, and pumping out the water from the new enclosures, and thus new fields were formed below sea level. Windmills, first used in the 15th century, assisted in pumping out the water from the newly reclaimed fields.[69] Between 1540 and 1615, around 80,000 hectares of agricultural land was created in the Netherlands, using a combination of sea dykes and drainage of inland waters.[70]

The reclaimed polder fields contain layers of peat, which formed along the coast as sea levels rose following the last ice age. The peat formed in a freshwater environment, where water draining from the rivers was backed up by natural coastal barriers such as sand bars. As the sea level rose, so did the groundwater levels, which also contributed to peat growth. After the 3rd century AD, the peat bogs were then flooded by sea water, which infused the peat with salt. This meant that peat could not only be used for fuel, but also for salt extraction, which became a major industry in the Middle Ages.[71]

In the Google Earth photo (Figure 2.6), you can see the canals and waterways between the fields. These canals are in fact early peat cuttings. In some places, lakes were formed where the peat cutting was extensive – like smaller versions of the Norfolk Broads, which were also created by medieval peat

Figure 2.6: Satellite image of Waterland, Noord Holland, The Netherlands. (Photo © Google Earth)

cuttings. You can see the small lake in the photo, just to the west of the village, but there are also much larger lakes and waterways. The peat cuttings had the effect of creating these low-lying fields that are surrounded by water, and thus created a landscape not all that different from the wetland fields of South America.

The Dutch created – seemingly by accident – pasture that is close to the water table, and the proximity to water prevents the land from freezing. In this respect, these low-lying fields experience some of the benefits of the purpose-built water meadows in the UK. The air photograph shows a protected area of land called Waterland, which is just north of Amsterdam, and this area is unusual in that the traditional, pre-industrial landscape has been preserved. Figure 2.7 shows the Waterland landscape from the ground, and the different wildflowers show something of the species diversity in the rich pasture. In much of the Netherlands the land has been reorganised so it can be farmed mechanically, and much of the biodiversity has been lost. Even so, the close-set drainage canals are a feature of the Dutch landscape throughout the low-lying areas of the country.

Coastal wetlands in the Netherlands: a prehistoric adaptation

There is also an Iron Age adaptation to flood risk that developed along the coast of the North Sea in what is now the Netherlands and northern Germany. The Iron Age farmers inland were growing crops and raising domestic livestock in a mixed economy, but the people along the coast were living in a more dynamic, changeable environment. They had to contend with high tides and storms coming from the sea, and also with floods coming from the rivers that empty into the North Sea.

Figure 2.7: Waterland from the ground. (Photo © Jerry Bond)

Early settlers on this land adapted by building on the high ground of levees (raised river banks), but as the sea level rose, the settlements had to be raised even higher. Artificial mounds (called 'terpen' in Dutch, or 'terps' by English speakers) were made of stacked turves and settlement rubbish, together with large quantities of animal manure.[72] The first ones were built in the 6th century BC, often originally being built to raise the level of a single farmstead. Over time, the many mounds built for individual farmsteads tended to merge into larger platforms supporting several houses and their barns. In the Roman period (*c.* 12 BC to AD 406) the farmsteads on the coastal terps expanded into larger settlements, with villages of 20–30 houses being established by the 2nd and 3rd centuries AD.[73] The mounds were used not just for houses and livestock, but also kept arable land securely above sea level, so that the crops were not harmed by floods or by salt. Terpen range in size from just a few hectares to 16 hectares at Ezinge, and are up to 9 m high.[74] A number of medieval villages are located on top of Iron Age terpen; the church was typically built at the highest point (Figures 2.8 and 2.9).

There is a vivid description of the Iron Age coastal Chauci tribe by Pliny the Elder, who took part in the conquest of the Germanic tribes in AD 47. He describes the Chauci as 'a wretched race living in a vast tract of land, invaded twice each day and night by the overflowing waves of the ocean', and he goes on to suggest that they were so impoverished, cold, half-starved and wretched that they would be better off

Figure 2.8: Church on an Iron Age terp, the Netherlands. (Photo © Jerry Bond)

being conquered by the Romans, who would at least introduce some of the comforts of civilisation:

> They fashion the mud... with their hands, and drying it by the help of the winds more than the sun, cook their food by its aid, and so warm their entrails, frozen as they are by the northern blasts; their only drink, too, is rainwater, which they collect in holes dug at the entrance of their abodes; and yet these nations, if this very day they were vanquished by the Roman people, would exclaim against being forced to slavery! Be it so, then – fortune is most kind to many, just when she means to punish them.[75]

It is interesting that he also notes that they lived only by fishing and did not farm, because excavations along the coast show that the Iron Age people of these wetlands had sizeable house-byres, with animal stalls at one end and the dwelling space at the other. The site of Feddersen Wierde, located in the wetlands of Northern Germany, is said to be typical.[76] It was first settled in the late 1st century BC, and it had an animal bone assemblage that was almost entirely made up of domesticated animals, with no evidence for fishing even though the site was right on the North Sea. Sixty thousand animal bones were identified, half of them from cattle and a quarter from sheep or goats; the remainder included pigs, horses and a few dogs – but only 1% of the animal bone was from wild animals. There were fish all around them, but no evidence for fishing. These people were living what was essentially a dryland life, but in a wetland region.

Figure 2.9: The edge of a terp mound. (Photo © Jerry Bond)

This may be because they were trading with people from the inland regions, where farmers practiced a mixed agriculture of crops and domestic livestock. At Feddersen Wierde there was just one granary per building, whereas the villages on the sandy inland region had many of these distinctive structures associated with each farmhouse.[77]

By the 1st century BC there was byre space for 100 animals at Feddersen Wierde, and by the 3rd century AD there was space for 450 animals. This means that the people would have had milk, meat, wool and traction animals. The focus on raising livestock in this area makes sense, given the increasing opportunities for trade that developed over the course of the Iron Age. The first towns in Europe – called 'oppida' by the Romans – were centres for trade, producing specialist crafts and minting coins before the arrival of the Romans. People were beginning to move away from a subsistence-based society (*i.e.* one in which individuals produce everything that they need for themselves) to a more complex way of life in which people specialised and sold their produce. The Roman conquest accelerated this process, and large numbers of Roman coins found in the terps indicate that the people in this isolated region were trading with the Romans – although they may have been simply bartering on a local level.[78] Feddersen Wiede was clearly part of this wider economic network, importing Roman goods such as the fine (but mass produced) Samian ware pottery, glass artefacts and fibulae (Roman brooches). Feddersen Wierde is but one example, and it should be emphasised that there were many such raised settlements in the region around Bremerhaven.

Sea level rise and land reclamation

Dutch land drainage and reclamation created the polders that characterise much of the Dutch landscape. The sea level – which has been rising since the last ice age – continues to rise. Peat extraction, which has created the dense canals between small fields, has also lowered the land level, but a further problem has been that when peat is exposed, it oxidises, shrinks, dries out, and is blown away by the wind.

In the Cambridgeshire fens, farmers have faced the same problem of peat wastage. Nevertheless, they have persisted in carrying out arable agriculture on this very rich farmland, despite the wind erosion that blows the dry peat into great clouds of dust. Studies of peat loss in Cambridgeshire began in the 19th century, and a famous relic of the early studies is the Holme Fen post, which was sunken into the fen peat until its top was flush with the land surface. This took place in order to monitor the changes to the land surface that were expected to take place when the new Whittlesey Mere was drained. In the first ten years of the experiment the land level dropped by 1.5 m, with another 1 m lost in the following ten years. Another 1.2 metres were lost in the next 100 years, and the post is now raised up on a concrete plinth in order to maintain the top at the level of the original land surface.

There is a cycle that occurs when peatlands are drained: the water is pumped out, the peat dries out, and the land surface drops and is subject to renewed flooding. In both England and the Netherlands, windmills accelerated the capacity for drainage –

but the peat levels kept dropping. Steam pumps replaced wind – and the land dropped still further. Diesel replaced steam, and more land eroded. The lower the land level, the more difficult it is to prevent flooding, and the more subject it becomes to potentially catastrophic flooding.

Catastrophic flooding

The Netherlands are essentially one large river delta, and floods are threatening the country from both land and sea. When rising sea levels meet with rising rivers, vast areas of this low-lying area are threatened. In 1953, a major storm over the North Sea caused catastrophic flooding and the deaths of nearly 2200 people. Three hundred and six were killed in Eastern England, and 19 were killed in Scotland, while 22 people died in Belgium. The worst of the destruction fell on the Netherlands, where 1836 people were killed when the dikes gave way and vast areas of land, totalling about 340,000 acres (137,593 hectares) was flooded. About 100,000 people had to be evacuated in the Netherlands, while 32,000 people were evacuated in England.[79]

The Dutch government improved the country's flood defences after the Great Flood of 1953, but the sea level is rising and the land is subsiding, and climate scientists are predicting more frequent thunderstorms, which will also be more severe. Sixty per cent of the Netherlands is below sea level, while the population is also increasing, and the government has therefore put together a new approach which combines old and new solutions – including some that are in fact prehistoric. In addition to strengthening the existing dikes and installing extra pumps – which are medieval but nonetheless effective technologies – the government is implementing a programme called 'Ruimte voor de Rivier' (Room for the River), a major project that involves reinstating large areas of floodplain, re-flooding polders and creating extra flood channels to prevent freshwater flooding inland.[80] The programme is actually made up of 34 sub-projects, many of which have already been completed.

Rivers need room to flood, and as sea level and groundwater levels rise, land surfaces are sinking and storms are growing more intense. In light of these changes, it is no longer enough to keep raising artificial levees along the river banks. The new programme is addressing the problem by working with nature instead of against it; it is allowing rivers to flood, but in a controlled manner. Ancillary channels have been built as part of the overall programme, so that there are managed overflow channels to prevent flooding in periods of heavy river flow.

In another local project, carried out in 2015, the River Maas was widened at Overdiepse Polder.[81] In the process, the dyke there was lowered so that the polder can be flooded in times of need – thus storing excess water and protecting important towns and villages downstream. To me as an archaeologist, however, the most interesting thing about this particular project is that the government has built eight new terps – thus using an Iron Age solution to work in a uniquely symbiotic way with 21st century flooding. The project has not been without pain, in that some old farms were demolished and re-established on the new terps, but it seems like a sensible, long-term solution. Terps have also been built on the River Dventer, including one for a horse farm. The farms – like the Iron Age villages – are now out of harm's way.

The re-establishment of floodplains does not mean that these areas of land are now single-purpose – far from it. About 20,000 hectares of floodplain will be used as pasture for cattle in times of low flood risk, which not only feeds the cattle, but also keeps the grass short.[82] The River Authorities regard this as an added benefit, because short grass creates 'low hydraulic roughness', which allows the water to drain away more rapidly.[83]

In other areas, 'retention polders' are created which are managed simply as wetlands. This is a blessing for biodiversity, because about 16% of the highly industrialised Dutch landscape is wetland of international importance.[84] These new wetland areas are already attracting wildlife and wildfowl, and they are also attracting walkers, cyclists, birdwatchers and parents with children. However, there is yet another benefit: wetlands emit 20–25% of global methane, but they store 20–30% of the organic carbon that is stored in soils around the world.[85] On balance, the positive effects of carbon storage in wetland soils and peats far outweighs the negative effects of methane emissions.[86] Wetlands are important for flood prevention, but also for their role in preventing further climate change.

Notes

1   Mitsch and Gosselink 2015.
2   Yellen *et al.* 1995.
3   Price 1991.
4   Rick *et al.* 2001.
5   Coles and Coles 1989, 144 and 146; Van de Noort and O'Sullivan 2006, 95.
6   See all references to Erickson and Denevan.
7   Parsons *et al.* 1985.
8   Coe 1964.
9   Outerbridge 1987.
10 Outerbridge 1987.
11 Outerbridge 1987.
12 Werner 1992.
13 Werner 1992.
14 Werner 1992.
15 Wilken 1985.
16 Parsons 1991, 37; Denevan 2001, 235.
17 Outerbridge 1987.
18 Werner 1992; Popper 2000.
19 Wilken 1985.
20 Wilken 1985.
21 Wilken 1985.
22 Parsons *et al.* 1985.
23 Parsons *et al.* 1985.
24 Parsons *et al.* 1985; Werner 1992.

25 Werner 1992.
26 Coe 1964.
27 Coe 1964.
28 Coe 1964.
29 Outerbridge 1987.
30 Outerbridge 1987.
31 Outerbridge 1987.
32 Outerbridge 1987.
33 Outerbridge 1987.
34 Werner 1992.
35 Wirth 1997.
36 Meggers 1954; 1971; 1957.
37 Meggers 1954.
38 Meggers 1971.
39 Erickson 2014.
40 Denevan 2001, 239.
41 Denevan 2001, 246.
42 Denevan 2001, 248; 1992.
43 Denevan 2001, 250.
44 Erickson 1995.
45 Boixadera *et al.* 2003.
46 Denevan 2001, 252; Erickson 1995.
47 Erickson 1988.
48 Erickson 1988; Denevan 2001, 258.
49 Erickson 1987; 1992.
50 Erickson 1988.
51 Erickson 1988.
52 Erickson 1988.
53 Erickson 2003.
54 Erickson 2003.
55 Erickson 2003.
56 Pretty and Shah 1997.
57 Allen and Robinson 1993, 145.
58 Allen and Robinson 1993, 145.
59 Cook *et al.* 2004.
60 Oxford English Dictionary.
61 Rackham 1986, 333.
62 Rackham 1986, 334.
63 Rackham 1986, 337.
64 Williamson 2007.
65 Rackham 1986, 338.
66 Williamson and Cook 2007.
67 Rackham 1986, 339.
68 Williamson 2007.
69 Ciriacono 2006, 161.
70 Ciriacono 2006, 176.
71 de Kraker 2015.
72 Boersma 2005.
73 Fokkens 1998, 131.

74 Boersma 2005.
75 Coles and Coles 1989, 42.
76 Parker 1965.
77 Fokkens 1998, 131.
78 Fokkens 1998, 131.
79 Hall 2013.
80 https://www.ruimtevoorderivier.nl/english/ Accessed 9 November 2018.
81 https://www.ruimtevoorderivier.nl/river-widening-overdiepse-polder/ Accessed 9 November 2018.
82 Brouwer et. al. 2001.
83 Smits *et al.* 2001.
84 Best *et al.* 1993.
85 Mitsch and Gosselink 2015.
86 Mitsch and Gosselink 2015.

3

Farming the desert

In the desert, people have no choice but to be wise.

Uzi Avner[1]

The Negev desert

In 1870 an English explorer, Edward Henry Palmer, travelled by camel along the presumed route of the Exodus, across the Sinai desert and then north into Jordan. Along the way he aroused much suspicion among the Bedouin tribesmen, who accused him of using his surveying equipment to prevent the rains from falling. In fact, he was surveying the ruins left by the ancient farmers who in antiquity had once managed to live in this parched desert landscape by harvesting the rain. He described a number of ruinous ancient settlements, including cairns, houses and cisterns as well as agricultural installations such as stone walls and dams, 'all of which bespeak a former state of fertility and industry'.[2] He suggested that the ruins must have been left by a 'more civilised people than those who now inhabit the country'.

Palmer is possibly the first westerner to record one of the most interesting features of the Negev desert:

> It is a noteworthy fact that among the most striking characteristics of the Negeb [sic] are miles of hill-sides and valleys covered with the small stone-heaps formed by sweeping together in regular swathes, the flints which strew the ground[3]... These at first considerably puzzled us, as they were evidently artificially made and intended for some agricultural purpose, but we could not conceive what plants had been grown on such dry and barren ground. Here again, Arab tradition came to our aid, and the name teleilat-el-'anab, 'grape mounds', solved the difficulty. These sunny

slopes, if well tended, with such supplies of water and agricultural appliances as the inhabitants... must have possessed, would have been admirably adapted to the growth of grapes, and the black flinty surface would radiate the solar heat, while these little mounds would allow the vines to trail along them and would still keep the clusters off the ground.[4]

All of this begs the question of how people managed to grow crops in this dry and desolate place. Has the climate changed? Was it rainier in the past?

In 1954 a team of scientists got together to study the ancient remains in the Negev desert.[5] The team included archaeologists, botanists, geologists, geographers and agronomists, and their initial findings confirmed what Palmer discovered in 1870, and much more. They found the ruins of six ancient towns, and the remains of stone walls and cairns covering *hundreds of thousands* of hectares. The research team used aerial photographs to map the extent of the walls in the desert, which has the advantage of being a more rapid and thorough method of investigation than ground survey. They took a number of their own photographs, relying on a Piper Cub because with this type of plane you can lean out of the open side, strapping yourself in with a harness. After examining their photographs, the team then went and mapped the most important areas on the ground.

When they began to map the pattern of the stone walls, they realised that there were several different types of systems with distinct uses, but all were built to manage the scarce water resources. Rainfall in the northern Negev is between 75–100 mm per year, which is regarded as true desert conditions and not just semi-arid land. When it rains, it rains in sudden, torrential cloudbursts which can cause severe erosion and gullying of the hillsides. The deep canyons that form are called *wadis* in Arabic, and after the rains they are subject to sudden flash floods, made worse by the nature of the fine, silty soils of the region. These soils have been described as being 'like face powder', and when it rains a crust is formed on the surface which is almost completely impermeable (Figure 3.1). The water flows off it instead of sinking in.

The early farmers, recognising this pattern of events, built stone dams across the wadis called check dams because they check the flow of the water. They stem the violence of the sudden flash floods, taming the water so it pools up behind the dam and trickles down into the next field, and the next, and so on down the staircase of fields[6] (Figures 3.2 and 3.3). When the floodwater pools up behind the dams it releases the accumulated sediment, which adds nutrients and also increases the depth of the soil until, gradually, a good, rich alluvial soil is formed in a series of small fields.[7] Trees and shrubs grow along the walls because of the deep, moist soil, and their roots are still helping to keep the walls in place several thousand years after their construction. Some of the ancient fields are still farmed today by the Bedouin.[8]

The check dams are easily understood, but what of Palmer's teleilat-el-'anab? It is possible that they really were used for viticulture, if the soils in the past were deeper and less saline than they are today? It has been suggested that the stones attract the dew, which condenses on the stones and then infiltrates down into the soil.[9] Another

Figure 3.1: Fine silt crust on the soil surface of the Negev desert near Avdat. (Photo © the author)

Figure 3.2: Aerial view of runoff farms in the Negev. (Photo © Google Earth)

suggestion is that they were cleared in order to increase the erosion of the silt soil, so that more sediment would build up behind the wadi check dams.[10] There was at one point some suggestion that they might be a natural feature, perhaps created by freezing and thawing.[11] Evenari and his team argued that their purpose must be to encourage the water to run off the soil surface more easily, so it could be collected and channelled into the fields. If water infiltrated around each stone on the soil surface, then a lot of the scant rainfall would be lost.

Evenari's suggestion was supported by research in the 1990s, which demonstrated that the volume of runoff from a slope with the surface stones removed can increase by up to 245%, compared to the natural, stony desert surface.[12] The experiment worked most effectively for short bursts of rain, which are the most common type of storm in the area. The runoff from the higher slopes was channelled by stone walls down into the wadis, with a careful system of sluices to ensure that each field got a share of the water.

Figure 3.3: Terraced wadis at Avdat. (Photo © the author)

Some wadis remained as individual farms, but larger units were also created by building walls around several wadis to create a larger farm. Thousands of these large farms have been found, each with a farmhouse with several rooms, a kitchen, storage facilities and deep cisterns for storing water underground, where it couldn't evaporate. Because of the harvested rainwater, the cultivated land on these farms would have received the equivalent of 300–500 mm of rain per year, although the actual level of rainfall was only around 100 mm.[13]

Fields created by check dams across the wadis have been dated to before 1200 BC at Wadi Faynan in Jordan,[14] but in the Negev they seem to be later. The simpler system of wadi check dams was thought by Evenari and his colleagues to date to the Iron Age Israelite period, before the 6th century BC, and although most subsequent surveys and analyses have cast doubt on the dating, there is one recent analysis that supports Evenari. Iron Age radiocarbon dates of 1052–971 BC and 828–759 BC were obtained for soil in a field in one of the simplest types of wadi system on another Negev desert farm, Horvat Haluquim, which is about 50 km south of Beer Sheva.[15] This suggests the fields at Horvat Haluquim were in use for many years, although it is possible that the use was discontinuous. The radiocarbon dated field is associated with an Iron Age II village (1000–500 BC).

The larger farm units and more complex systems of channelling water were thought by Evenari's team to originate in the Nabatean period, which dates from the 6th century BC until around AD 100, ending when the Nabateans came under Roman control. A number of stone troughs with Nabatean inscriptions were found, one of

Dating in the Levant

Bronze Age: 3300 BC–1200 BC

Israelite Period:
Iron Age I: 1200–1000 BC
Iron Age II: 1000–500 BC
(Iron Age IIa: 1000–925 BC)
(Iron Age IIb to IIc: 925–586 BC)
Iron Age III: 586–539 BC

Nabatean: earliest dates suggested are in the 6th century BC, ending around 100 AD

Babylonian and Persian: 586–332 BC
Hellenistic: 332–37 BC
Roman: 37 BC – AD 132
Byzantine: AD 324–638

Early Arab: AD 638–1099

which could be dated to the year AD 88–89 – but there was no direct link between the troughs and the runoff systems. Scholars today disagree with the original dating, but all agree that the systems were flourishing by the Byzantine period (AD 324–638).

The reason for the difficulties with the dating evidence is that archaeologists traditionally depended on pottery, which can have a large date range and which can also be displaced and re-deposited out of its original context – especially in an agricultural landscape. The Horvat Haluquim dates were obtained from radiocarbon, which is more precise than pottery, but now a new study has also been done using optically stimulated luminescence (OSL). This is a method of dating quartz grains and measuring the last time that they were exposed to the light, and a survey and OSL investigation published in 2013 dates the larger and more complex field systems to the 4th to 11th centuries AD.[16] This means they date from the Byzantine period and continued to function after the Arab Conquest of AD 634–640, but fell out of use towards the end of the Early Islamic period.

The development of the terrace systems from simple dams to elaborate, landscape-wide systems has been traced by means of excavation. One interesting development was found in a particularly large wadi, which had a check dam built across it that was later destroyed by flash floods. The farmer-engineers had attempted to manage a wadi that was simply too big, and it was destroyed by the volume of water and sediment that came crashing down the canyon. The amount of silt that had built up behind the walls was also a factor in overstretching the system – it's important to collect the fertile alluvium, but not so much of it that the walls break down. New stone walls were built using a different approach, leading the water out of the wadi and onto terraces which were built out of the direct flow of the stream. In the final phase, a stone aqueduct was built to carry the water from the wadi to the terraced fields. The early engineers had clearly learned from experience and revised the system until it was able to withstand local conditions.

In 1936 an American excavator called H. D. Colt was carrying out an excavation at Shivta, one of the six large towns in the Negev; we now know that it dates to the Byzantine and Early Islamic periods, and it is associated with rainwater harvesting agriculture. However, Colt's work had to be abandoned because a severe drought forced him off his site, so he went and excavated at another site, Nitzana, instead[17] (these were the days before licensing and regulation made this sort of thing impossible – archaeologists are no longer allowed to simply go and dig wherever we like without telling the authorities). His team found a number of papyrus documents, written in Greek towards the end of the Byzantine occupation of the region in the 6th and 7th centuries AD. The papyri include legal documents (including laws on water rights), a copy of Virgil, a Latin-Greek glossary of the *Aeneid*, the gospel according to St. John – and also records of land ownership and details of what was grown. They record grape vines, fig trees, date trees, almond trees, pomegranates, olives, wheat and barley – and they even recorded the yields. The stone-built wine and olive presses used for mass production and export are still to be found in the ruins of the towns of Avdat and Shivta (Figure 3.4).

Figure 3.4: Olive oil or wine press, Avdat. (Photo © Jerry Bond)

The environmentalists on Evenari's team believed that the climate in the region had not substantially changed in the last 2000 years, and they had the actual records of what was grown on the Negev farms. They were intrigued – and they decided to rebuild a farm to see how productive they could make the desert, using the runoff techniques of 2000 years ago. They wanted to test their theories on how the systems worked, to see how practical it would be to put such systems back into use, and they also wanted to study the natural history of the desert, to see how the plants and animals had adapted to the arid conditions.

They began with the fields surrounding the ruined city of Shivta, in 1958. The field walls there were exceptionally well preserved, so they could see exactly how to rebuild them, using silt to pack the spaces within the stone walls. By a lucky coincidence, the government tourist office began restoring the magnificent ruins of the town at about the same time, so the scientists had help from the labourers who were already on site. They had hoped to actually live in the farm buildings, but the site was a bit too far off the beaten track and it was too difficult to maintain a supply line, and also it was too dangerous. The experiment continued but there were some problems, including the crops being eaten by various passers-by.

In 1959 a second research station was built in the valley near the ruins of the desert town of Avdat (Figure 3.5). The experimental farm at Avdat was much larger than Shivta, with 2.6 hectares on 14 terraces. This was not the whole of the original farm,

Figure 3.5: The ruins of Avdat. (Photo © Jerry Bond)

but they put together enough for the initial experiments to be carried out. At Shivta the reconstruction followed the original pre-industrial methods, but at Avdat they used concrete on the walls, laid modern pipes and installed a pump in the cistern.

The first consideration was what they should grow. The Nitzana papyri provided a record of what had grown in the region in the past, but the team was sceptical at first. They were particularly uncertain about the viability of growing fruit trees in the desert, but they were encouraged by a comment recorded by St. Jerome (AD 340–420) on the excellence of the wine at the nearby town of Halutza. This Nabatean to Byzantine town was 12 miles south of Beer Sheva and is known today for its olive oil, which is regarded as one of the best tasting (and with the highest vitamin E content). Today the land there is irrigated with water from a deep aquifer rather than runoff, but the team decided to go ahead and take the risk.

The team chose mainly trees that were known to have grown in the Negev in ancient times, but they also grew trees that were suited to the climate or could adapt to it. This proved a great success; fig trees, vines, pomegranates and olives did well, and carob was even more successful. The almond and pistachio trees were very productive and the output was equal to that of the modern, irrigated Israeli farms. What's more, a calculation of the amount of water required by some of their crops showed that most modern farmers were over-irrigating, wasting a considerable amount of water.[18] It is now well established that modern irrigation tends to over-water crops.[19]

The peach and apricot trees provided an interesting surprise: living on 2700 cubic meters of water per hectare, they were not as productive as the more heavily watered trees found in modern systems (7000 m³ of water for peaches, 5500 m³ for apricots), but they produced the best fruit that any of the team and their families had ever tasted. When trees receive more water, their fruit becomes more watery and the sugars are diluted. A similar effect was found in some of the medicinal plants that they grew, which had a much higher concentration of the active compounds than is found in more heavily irrigated plants.[20]

The farm experiments would probably not have been undertaken if the scientists had known that Israel was about to enter the most extreme drought ever recorded in the Negev. Over the next three years the crops died all across the northern Negev and the southern coastal plain of Israel – and yet, Avdat produced two tonnes of wheat per ha and 2.7 tonnes of barley, the trees continued to grow, and the asparagus and artichokes thrived.

One interesting side effect of the drought was the diminution in crop pests that had otherwise been a serious nuisance. The animals, however, seem to have been a persistent problem, the scientists bewailing that, 'Our green fields attracted all the plant-eating population in the region and we sometimes imagined that there must be some kind of bush telegraph informing all the animals that they could save themselves from starvation by visiting the farms.' The farm had to be fenced in to protect the crops from hares, gazelles and porcupines. Raiding partridges were shot and eaten by the Bedouin, which made them a rather more welcome pest. The insect problem was addressed by the use of pesticides, which is a serious departure from the original farming methods. Inorganic fertilisers were also used, as well as organic manures.

Avdat and Shivta were small farms, but in the early 1970s another experimental farm was established in Wadi Mashash in order to find out whether runoff farming might be profitable if run on a larger scale. The team were particularly interested in creating improved pasture, given that the local population depended largely on their stock of sheep and goats. They also planted more orchards, mainly almond trees and also olives and pistachios. The pasture did very well despite there being only one runoff episode that year, while all around them the natural desert vegetation was in 'a deplorable state'. After the pasture was established, they bought a herd of sheep from their Bedouin neighbours. They chose an ancient Middle Eastern breed, the Awassi, because of its adaptation to desert conditions; the Israeli farmers tend to keep more productive breeds, but the Awassi are more appropriate for desert conditions because they need less water and food. The Awassi store fat in their tails instead of under the skin, so that their body heat dissipates more efficiently.[21]

The Bedouin goats are another impressive species, needing water only once every two to four days and eating very little.[22] Their bodies are actually 70–80% water, which is higher than any other mammal (dolphins, for instance, are only 37% water; people are 65%). The Bedouin goats also have extremely low metabolic rates, the lowest ever recorded in animals of their size, and this means that they – like the Awassi sheep – can maintain a low body temperature in the heat of the desert.

Cisterns

In 1914, just before the war, two British military intelligence agents were sent to map the Negev Desert. T. E. Lawrence (Lawrence of Arabia) and C. L. Woolley spent nearly two months mapping the region, while recording all of the archaeological ruins that they encountered. Their comments are worth recording:

> We believe and shall attempt to show that the prosperity of the Byzantine Age was wholly due to the conservation of the normal water supply and to improved agricultural methods. The towns and homesteads of the Byzantine Negeb [sic] relied entirely, as to some extent does modern Jerusalem, upon stored water.[23]

They visited the ruined city of Shivta (also called Esbeita) and noted that every house in the town had two cisterns on average:

> ... every street was graded down to a catch-pit; every courtyard and roof, even the flat ground outside the town, fed some underground store.

Woolley and Lawrence describe a cistern still in use by the Bedouin, which still held 'many feet of water' despite a two year drought; this cistern was still supplying water for all of the Bedouin sheep and goats for miles around.

The underground cisterns that retain the water are an important part of the rainwater harvesting system (Figure 3.6). There are hundreds of cisterns in the Negev,

Figure 3.6: Cistern at Masada, on the Dead Sea. (Photo © Jerry Bond)

Figure 3.7: Cistern at Avdat, with settling tank. (Photo © Jerry Bond)

mostly constructed to receive water from the smaller wadis. Evenari's team cleared out one of the cisterns at Shivta, and found that it supplied enough water for five or six families for a year, *and* their herds of sheep and goats. The Nabatean city of Petra, in Jordan, had huge municipal cisterns to collect and hold millions of gallons of rainwater. The insides were plastered so that they were waterproof, and because they were underground they kept the water cool and prevented it from evaporating. The water in these desert cisterns was usually channelled first into a small basin for settling out the silt in the runoff water, and clean, clear water then overflowed from the settling tank into the cistern itself (Figure 3.7).

Cisterns date back about 4000 years in Jordan, in areas with a comparable rainfall to the Negev. There is in fact a historical account written in Moabite (a Semitic language spoken in the Southern Levant) around 850 BC, recording the building of reservoirs and cisterns. This is written on the Meshe Stele or Moabite Stone, now in the Louvre but recovered, with difficulty, from Jordan in the 19th century.[24] It was found by a travelling German missionary, who told the German consul about his discovery. French diplomats in Jerusalem then heard of the stone, and they sought it out and attempted to buy it from the Bedouin. The Bedouin refused to sell, according to one account (written in 1871) because they regarded it as a sacred talisman, and to another because they thought they could make more money if they smashed it up and sold it in pieces. According to the latter story, the Arabs could see that there was a bidding war taking place between the Prussians and the French,

> ... so that its Arab possessors perceived that the stone was of great value, and conceived the happy idea of breaking it in pieces, in order, probably, to make more money by selling it in portions.[25]

Fortunately, a number of people involved in this fiasco had made partial plaster casts of the writing, and in time the fragments were mostly (although not all) recovered and re-assembled (it is now in the Louvre, and a copy is in the British Museum).

The stone was carved for King Mesha of Moab, which is located in today's Jordan, along the eastern side of the Dead Sea. King Mesha wished to record his conquests and building works, but among his many achievements, he arranged the building of reservoirs by Israelite prisoners, and also commanded his subjects to build household cisterns:

> I built Qerihoh: the wall of the parkland and the wall of the acropolis; and I built its gates, and I built its towers, and I built the king's house; and I made banks for the water reservoir inside the town; and there was no cistern inside the town, in Qerihoh, and I said to all the people:, 'make yourself each a cistern in his house'; and I dug the ditches for Qerihoh with prisoners of Israel.[26]

Excavators at Qerihoh, the capitol of Moab (the Kir Haraseth of the Bible, and today known as Dhiban or Dibon, Jordan), found 100 of these cisterns in excavations in the 1950s.[27] They used to be a common feature in the Middle East, and are found on most archaeological sites in the region. As recently as 1921, a census in Jerusalem recorded 7000 cisterns for collecting runoff water.[28] There are still some cisterns in use in Jordan, including ones which are thought to date to the Nabatean period.

Unfortunately, the cisterns that survive in Jordan and Israel today have mostly gone out of use and are now filled with sediment and rubbish. The water supply has been centralised, and is piped in from large reservoirs. In Jordan, this means that 92% of the annual rainfall simply evaporates.[29] This is an incredible waste. A survey commissioned by the International Center for Agricultural Research in the Dry Areas (ICARDA), a research centre in Syria, suggests that water harvesting could save 10–15% of the water currently being lost in Jordan (6.6 billion cubic meters).[30] This could increase the level of water in the aquifers, and they suggest it could double the annual agricultural production, as well as providing supplies for both domestic and industrial use.

In Jordan there was a project in the 1990s called 'Project Rainkeep', that restored ancient cisterns and dug new ones as part of a water conservation strategy. It was a partial success; most of the villagers who benefitted from the program said that the rainwater tasted better than piped water, but there were disagreements about the maintenance work on the cisterns and the cost-effectiveness of the project overall.[31] The original plan entailed solar powered pumps and drip irrigation, but in practice, it was left to the women to do the pumping by hand – which may have been part of the problem. The main problem seems to have been the top-down approach, with poor communication between the agency that restored the cisterns and the families that benefitted. The villagers themselves felt that the project was a success, however, and happily there are now further projects in progress that will help to rejuvenate the rainwater harvesting systems and water cisterns in Jordan.

Water harvesting and storage of rainwater is to be found all over the Middle East and North Africa, and there is growing interest in developing these ancient resources.[32] Each region has its own distinct traditions, and each has its own regional name for the water capture and storage features. In Morocco, underground rain-fed cisterns are called 'matfias', and the early ones are still in use, but new ones are also being built.[33] The traditional matfias had round cavities and small storage capacities of around 5 to 12 cubic metres, and farmers would own many such cisterns. Today, large, rectangular, concrete-lined matfias for public use are being built by the forward-thinking Moroccan government.

The viability of runoff farming

On a technical level, the redeveloped experimental farms in the Negev proved to be a great success, in that they were both resilient and productive. Their resilience is demonstrated by the crop performance during the drought of 1961–63, which destroyed the modern irrigated crops, and the productivity is demonstrated by the number of crops that were produced in quantities that were competitive with modern irrigated farms. Why, then, did they fall out of use? The answer seems to be political and economic, rather than environmental. The Byzantine Empire came under attack in the 7th century, first by the Persians and then by the newly Islamic Arabs. The cities of the Negev were abandoned, but interestingly the farms continued to flourish in the Early Islamic period, declining in the 9th or 10th century and finally going out of use in the 10th or 11th century, at a time when the climate had actually become more favourable.[34]

Evenari's team believed that runoff farming should be reintroduced in developing countries, and the idea caught the attention of agronomists from all over the world. A German team working in Afghanistan brought the idea of runoff farming to help the rural people there, and they developed 70,000 ha of farmland.[35] The idea also spread to Botswana, Niger, Upper Volta (now called Burkina Faso) and Kenya.[36]

The redevelopment of early systems has to work within the economic and social setting of the people who are to adopt it, and Evenari's team learned this through the first-hand experience of getting it wrong. At the experimental farm in Wadi Mashash, Evenari's team re-excavated an early well. It produced excellent drinking water, and the team installed a hand pump and offered to share it with the local Bedouin. The scientists spent a week helping them to operate the pump, then left the Bedouin free to use it at any time. After a few days, the tribesmen returned to their old waterhole, which was dirty, shallow, and which was used by lowering an old tin on a piece of rope. The scientists asked them why, when they have access to all this clean water, did they return to the muddy water hole? The Bedouin complained that the hand pump gave them blisters and was too much work, but the excavators suspected a cultural reason: typically, it is the women and children who collect water. By returning to the dirty waterhole, the men could relinquish responsibility and allow the women to do the work, sparing themselves effort and blisters. The scientists describe this as 'an interesting lesson in how difficult it can be to change age-old customs and habits'.[37] Today, there is such a wealth of similar stories that the 'top-down' approach to development has been largely rejected, and local people – who know their own landscape – are included in discussions rather than having systems set up for them without their input.[38]

The idea of using runoff farming in developed countries has not been seriously contemplated. Evenari's team found that their tractors got bogged down in the sticky wet silt, and camels were not strong enough to pull the heavy plough; in the end, they resorted to using a rototiller. (It is worth noting here that the Bedouin use camels to draw light-weight wooden ploughs, which are better suited to the conditions.) A report by the UN in 1967 argued that runoff farming on a commercial scale was unfeasible because the supply of water could never be predicted, and the soils of the flooded terraces are too variable, with widely differing moisture content – but since then the UN has become more enthusiastic about rainwater harvesting as an efficient method of farming in arid regions. Evenari's team found that they simply had to be adaptable, and to accept that they could not be in total control of what they grew. The floods might come early in the rainy season, when the temperatures are low, or they could come later, when the temperatures are high. The crops that they planted would have to vary, depending on when the rains came. Consequently, they planted winter field crops if the rains came early (winter wheat, barley, legumes and vegetables) and summer field crops if the rains came late (safflower, chick peas, cotton, sesame and sorghum).

Runoff farming is not suitable to all areas of the desert. Experiments by A. Yair in the 1980s showed that rocky slopes are better for collecting rain than those with more soil, probably because the rainwater tends to run down the rocky bits but sinks into the soil.[39] A lot of rainwater can be lost on slopes with a lot of colluvium (eroded soil from off a slope) at the base. The early farmers preferred to use the steepest slopes where the water was most likely to run off rather than sink in, and this explains why the runoff farms are concentrated in the steep Negev highlands rather than the rolling hills near Beer Sheva, which at first glance appear to be more suitable. The creation of the farms was also labour intensive, with the reconstruction of Shivta taking 2000 man days.[40] In the past, the building may have been done by family units, or it may have been done communally, with each family helping the others to build their farms.

In the 21st century the climate of opinion is changing, and the possibility of combining old and new techniques may make the prospect of runoff farming more appealing to small farmers. The experimental farms have certainly demonstrated that the farms can be productive, as productive per hectare as the most modern irrigated farms. They are resilient and continue producing during severe drought, and perhaps most importantly they are sustainable: they don't deplete the water table and they don't cause salinisation of the soil, a recurring problem where crops are over-irrigated in the desert (see inset box).

Salinisation

All rocks contain salt, and when rocks break down they release salt into the soils and groundwater. In temperate regions the salt is washed away in streams, but in arid regions it builds up in the groundwater. It is important that the water table in arid regions is kept well below the soil surface, so that it does not damage or kill the crops. Over-irrigation raises the level of the water table, and evaporation pulls the water up through the soil, where the salt can precipitate out onto the soil surface, forming a toxic crust.

The Anasazi

It is worth noting that the Anasazi in the American Southwest also had a system for collecting water in the desert. They used the flat-topped mesas as catchments for gathering rainwater, putting earth dams across the canyons to catch the water that cascaded down the gullies.[41] The water was then channelled through a series of canals into the gardens and fields. The system went out of use when the desert became too arid during a 60 year drought; we don't know for sure why the people left, but it seems probable that climate change made the region uninhabitable for farmers.[42]

Libya

In 1978 the Libyan leader, Muammar al Gaddafi, said in a public speech that 'if archaeology is to be practised at all, then at least let it be relevant to the needs of people today'.[43] If this seems a bit harsh, consider that Charles Clarke, the British Labour minister for education from 2002–04, made a similar point when he stated that education for its own sake was 'a bit dodgy'.[44] According to the Times Higher Education Supplement, Clarke also told a group at University College Worcester that 'I don't mind there being some medievalists around for ornamental purposes, but there is no reason for the state to pay for them', adding that the state should only pay for subjects that exhibit some 'clear usefulness'.[45]

Most educated people would argue that understanding the past is important for its own sake, but also that history has an inherent practical application in that 'Those who cannot remember the past are condemned to repeat it.'[46] It is a testament to the importance of history and archaeology that Colonel Gaddafi invited the United Nations Educational, Scientific and Cultural Organization (UNESCO) to investigate the ancient floodwater farming systems in the Libyan desert, with an eye to putting them back to work.[47] Teams from Britain and France worked together with Libyan colleagues to carry out extensive surveys in the Libyan desert, mapping the remains of stone walls, cisterns and settlements.

Libya has a narrow but fertile coastal zone, with fertile highland steppes immediately to the south. Further inland is a region of sparse grassland, the 'pre-desert', and beyond this is the Sahara Desert. During Roman times the coastal zone of North Africa was an important agricultural region, but the pre-desert was also adapted for agriculture using rainwater harvesting techniques. Large, unenclosed farms were built in the 1st–3rd centuries AD, and in the 3rd and 4th centuries more substantial, fortified farmsteads were constructed.[48] The rainwater collection systems may not have been as elaborate as those in the Negev, but many elements were present. An extensive system of walls diverted water from the plateaus and hillsides into cisterns, often with sediment traps to keep the water clean. Stone walls on the wadi floors caught the water and allowed it to sink slowly into the alluvium that formed the arable land, also trapping the rich sediment that was swept down off the hillsides. Many of the stone walls contain gaps which allowed the water into a stone-reinforced area just downstream, so the force of the water wouldn't break down the walls or scour away the soil.

The walls were built by a wide range of methods, with some very substantial structures built to take the full force of the water, and others built to regulate a more gentle flow. Channels were cut into the rock in some places, and traces of cement indicate that every effort was made to ensure that none of the water escaped. There is even some suggestion that surface stones may have been removed to promote runoff, although the evidence is not as clear as in the Negev.

The Romano-Libyans grew wheat, barley, olives, grape vines, figs, almonds, pomegranates, dates, pistachios, lentils, peas and vegetables in the pre-desert. They kept sheep, goats, cattle, pigs and camels and sometimes chickens, and the domestic animals seem to have been carefully managed so that they didn't trample the crops or overgraze the wider landscape. This is important, because overgrazing causes severe soil erosion and has been a key problem in later times.

Inscriptions found in the settlements suggest that the people who built these runoff farms were local Libyans who had been absorbed into the Roman empire, rather than invaders who supplanted the natives[49]. Over time, fortified settlements were built with olive presses and large plaster lined cisterns, which may represent the increased power of a local elite who controlled the system.[50]

The farms went slowly out of use when the Roman Empire collapsed and the international trade system broke down, but some of the farms are still operating today. Many of the Romano-Libyan cisterns are still functioning, and huge numbers of the stone walls along the wadis are still standing. The excavation team emphasise that runoff farming is a sustainable enterprise – we have substantial evidence to show that it was successfully practiced for almost 2000 years – but some provision would have to be made for potential crop failures. Experimental farms need to be built, like the ones in the Negev, before we can safely say that the systems will prove viable – but given the evidence of long-term sustainability, with no serious soil erosion and no salinisation of the soil, there is every reason to believe that the system could work once again in Libya, as it is beginning to work in other places around the world.

Qanats

One of the most astounding feats of engineering is the Middle Eastern development of underground canals, called qanats (the Semitic word), foggara (in Egypt, Libya and Algeria) or karez (in Iran, Afghanistan, Pakistan and China) (Figure 3.8). The canals brought water from aquifers in the mountains out onto the plains, enabling people to establish settlements on land with no local water source. The system was invented before 1000 BC in Persia (the region that is now Iran),[51] and the idea spread along the Silk Route to China, as well as to Afghanistan, Pakistan, Iraq, Syria, the Arabian Peninsula and the Levant. Qanats were built in North Africa, where they may have been introduced by the Romans, and subsequently Arab engineers brought the idea to Spain, and the Spanish brought the idea to South America.[52]

We have records from many different periods describing how the qanats were built. The Roman historian of technology, Vitruvius, described them in the 1st century BC, and there is a Persian treatise on them that was written in the 9th century AD.[53] The first step to building a qanat was the surveying work, in which the engineer would look for indications of underground water on a mountain slope. They would start by finding an alluvial fan and then they looked for any sign of water, such as a visible spring or a change in the vegetation. They then dug the first well, the mother well, and then the tunnel below ground. The technique remained more or less unchanged for thousands of years, and in the 1960s a modern observer described how two men excavate the well

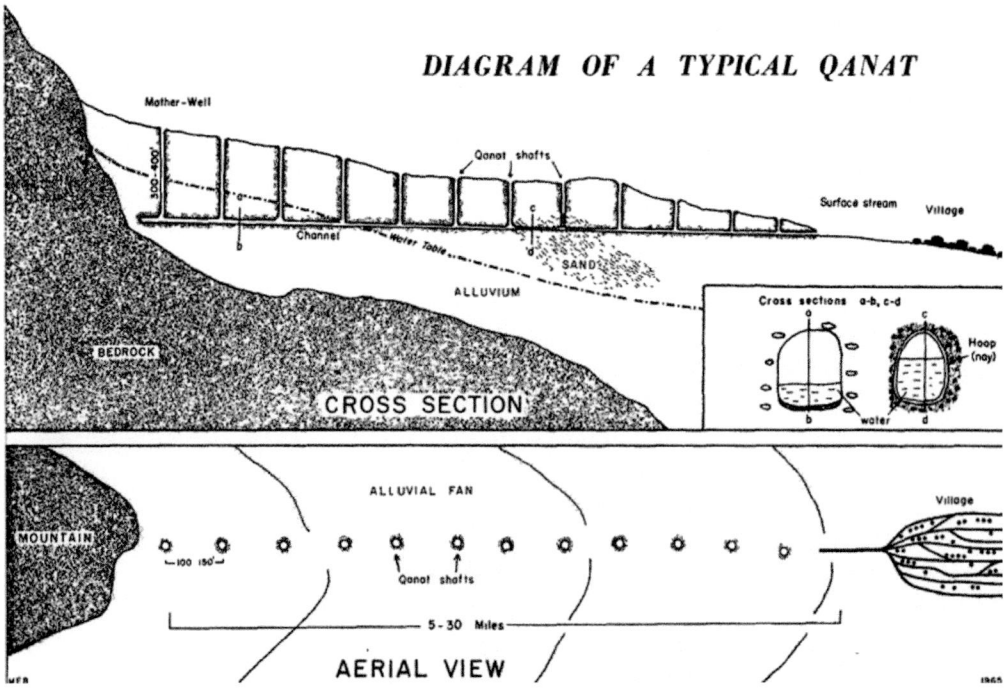

Figure 3.8: Diagram of a qanat (from Wulff 1968).

Figure 3.9: Aerial view of qanat wells. (Photo © S. H. Rashedi)

shaft, setting up a windlass to haul up the soil which is piled around the well to protect the entrance.[54] The course of the canal is surveyed, making sure that the gradient is very slight so that the water will flow gently and not destructively into the settlement and fields at the other end. The excavators begin digging the canal at the outflow end, working their way in a straight line back towards the mother well by sighting along two oil lamps. A chain of well shafts is excavated all along the course of the route for ventilation and to remove the soil; these are kept open and are used to repair and maintain the qanat (Figure 3.9). In recent and historical times qanat excavation and maintenance was undertaken by specialists who pass the skills down from father to son.

The Assyrian King Sargon II (722–705 BC) wrote that he had learned how to build qanats during his military campaign in Northern Persia, the area that is now Armenia.[55] His son, Sennacherib, built qanats to supply water to his palace in Ninevah, around 700 BC. Qanats brought water out into otherwise arid land, enabling new agricultural settlements to be established, and thousands of new farms and villages were built under the Archaemenid Empire (a phase of the Persian Empire, dating to 559–330 BC). The Greeks used this technology to supply Athens with water, and the Romans and Byzantines spread the idea to Jordan and Syria[56] (in Syria they are called 'Roman canals').

In the 1950s Iran still had more than 273,000 km of underground canals, supplying 75% of the country's water, and the city of Tehran was entirely supplied by qanats.[57] The oasis city of Turpan, in China, still had 5000 km of qanats (or karez) systems in the 1950s before the installation of deep wells and power-driven pumps. Unfortunately, the introduction of deep tube wells and pumps has replaced many of the qanats in countries across the Middle East and Asia. The problem with deep wells is that they deplete the aquifers, and new wells have to be dug deeper.[58] When the ground water level drops, shallow wells and springs also dry up, which causes serious problems for the people who depend on them – usually the poorest people. Another reason that qanats go out of use is that people regard them as too labour intensive to maintain, it being easier to rely on wells with mechanical pumps – but it is worth noting that tube wells have a finite lifespan, whereas a well-maintained qanat lasts for hundreds of years.

The advantage of the qanat systems is that they do not affect the water table, so they are sustainable. Qanats have provided water for large populations across the arid regions for thousands of years; they require no modern technology (so they are appropriate for developing countries), they cause no pollution, and they are maintained by the people who depend on them, so it is a reasonably equitable system.

It has been argued that qanats are a great concept but that they waste a lot of water because they are free flowing.[59] It has also been said that they are too dangerous to be used in modern times, and it is true that traditionally they involved child labour.[60] I would agree that the key disadvantage of qanats is the safety issue, but where modern qanats are being built they are being reinforced with concrete or bricks to improve stability. In Japan, where qanats were introduced in the 16th or 17th century, some of the qanats (called mambos or manbos, from the medieval Japanese word for 'tunnel') have been lined with large concrete pipes, which has drastically reduced the amount of work required to keep them clear, and has also reduced the amount of water lost to seepage as well as making them much safer.

Cheap reservoirs and cisterns can be built to conserve water in the settlements and farms, and the use of child labour is not essential to their maintenance, any more than slave labour is required to rebuild the chinampas of Mexico. Leaving issues of child and slave labour aside, it is important for each generation to pass on the knowledge and engineering skills of qanat building to the next generation, a process noted by Daniel Hillel in the 1950s.[61]

One region that stands to benefit from the redevelopment of the qanats is Oman, where around 3017 of these underground canals are still in use.[62] The system of qanats was brought to Oman in or before the 1st century AD, and they are still an important source of water for the mountain villages and oases today.[63] In 1982 it was estimated that qanats delivered 71.3% of Oman's water, irrigating 55% of the arable land.[64] It was also estimated that 80% of the country's qanats were in need of repair, but fortunately the government recognised the important contribution that they made to the economy and invested a great deal of money and materials towards their repair and upkeep. This is significant, because in many countries qanats are regarded as 'primitive' and governments are eager to bring in new, western style technology.

A key problem in Oman was the amount of food that was being imported, as the prices for imports were increasing exponentially. The rural economy was in difficulties because the farmers weren't earning enough money growing fruit and vegetables (which made up about a quarter of the food imports), and the young men were leaving the countryside for better paying jobs in the city. A programme to increase the efficiency of the qanat system was set up under the direction of Roderick Dutton, the director of the Centre for Overseas Research and Development at the University of Durham, and in his research[65] he found ways of addressing all of the concerns expressed by qanat critics.

Dutton identified a number of problems, including three which were of particular importance:

1. The water from the qanats flowed continuously, so a lot was wasted.
2. The way that the water was circulated around the village meant that each area only got its share of water every 6–8 days.
3. Many arable areas were abandoned because they were traditionally used for growing wheat, which was no longer profitable.

He then pointed out how these problems can be addressed, firstly by lining the canals so that water isn't lost, and by making them narrower and deeper to reduce evaporation. Cisterns need to be built on every farm, and the cheapest and most efficient way to pump the water from cisterns to the fields is by solar pump – which can also automatically regulate a system of sluices. The cisterns and sluices can be used to regulate the flow of water around the system, so that each farm will receive water every day. This would mean that more water-demanding but more profitable fruits and vegetables could be grown. The farmers make a greater profit, labour costs are reduced, and the villagers remain in charge of their own irrigation system. The labour costs are an important factor, as there are actually labour shortages in the villages as the young men go to look for work in the cities.

Dutton's proposal would enable the ancient system to continue functioning, but with a degree of mechanisation. The mechanisation can be done using renewable solar energy, the use of water can be made much more efficient and new cash crops can be grown. There is no reason why the qanat economy cannot be revitalised in many more regions – the key is to look at local factors, such as the availability and cost of labour, the economic potential, the degree of government support and the degree of local willingness to maintain the systems.

Many other countries are also beginning to recognise the advantages of using qanats. Between 1998 and 2003, 7858 qanats were rebuilt in Iran – more than 25% of the country's estimated 30,000.[66] In Pakistan, 112 have been rebuilt in the province of Baluchistan, together with storage areas for the water so it doesn't drain away. Qanats are increasingly becoming a part of government planning in arid areas, and planning laws are being introduced that protect the water flow – for instance, in Afghanistan there are now restrictions on the sinking of deep wells that would deplete the aquifers and damage water flow in the qanats. In Northern Iraq, UNESCO has helped to restore qanats, cleaning out the tunnels but also lining them so that they are safer and easier to clean in the future. Training programs are also underway in many regions, so that local people can learn the skills to maintain and manage the qanats. Retaining tanks are being added to save water; linings are being added to improve safety and ease cleaning, and caps are being placed over the shafts to prevent children and animals from falling in.

Rock mulching

Rock or lithic mulching is the practice of spreading stones or volcanic ash onto the surface of the soil to protect it from evaporation. It can be used in fields or gardens, and can be raked up into mounds or ridges, or just spread over the soil. In addition to reducing evaporation it increases infiltration of water into the soil – as we saw in the Negev. Rain sinks into the ground around the stones, and whereas in the Negev this principle was applied in order to increase runoff water for harvesting, in this instance it is simply to harvest water *in situ*, in the fields. By reducing evaporation, rock mulches also reduce salinisation of the soil.[67] The rocks raise the soil temperature during cold nights, radiating back the heat that was accumulated during the day. Rock mulching roughens the ground surface enough to create a more turbulent air stream over the soil, which also slightly reduces the temperature in the day and increases it at night. The increase in wind turbulence reduces the wind velocity, and so reduces wind erosion as well. Regulation of the temperature means that the growing season is extended, and the warmer night-time soil temperature enables plant roots to take up water and nutrients more rapidly, speeding up germination and healthy growth and producing higher yields.[68] The root systems are better developed under this system, and more able to withstand drought. The drawback to the system of rock mulching is that it hinders the application of fertiliser, but this isn't an insurmountable problem. Organic matter, including crop residues, can't return to the soil when there is a covering of rocks, so the stones are periodically removed, manures are applied, and then the stones are replaced.

In recent historical times rock mulching was used in China from at least as early as 1800 (but possibly much earlier) until 1948, the most recent reference. It was used on farms in the United States in the 1930s, 40s and 60s, and today rocks are still piled around the trunks of fruit trees and pecan trees in all different regions of the country. Olive trees are similarly protected in Syria. In the Canary Islands, a volcano that exploded in 1740 covered the countryside of Lanzarote with a gravel-sized cinder called lapilli. This was such an effective rock mulch that farmers have searched it out and spread it onto their fields ever since, even if it has to be transported from other local volcanoes.

We can also detect rock mulching in archaeological contexts. Over 400 ha of fields were mulched by gravel on South Island, New Zealand, where from AD 1200–1800 the Maori grew sweet potatoes and possibly maize.[69] There are also vast areas of stone strips, mounds and ridges in northwest Argentina, which date to at least AD 300–700 but possibly much earlier, and which are thought to have been used until about AD 1500.[70] Prehistoric rock mulching is mainly known from the south-western United States, where the Anasazi Indians in New Mexico created extensive pebble mulch fields around their settlements between AD 1350–1450, covering at least 7 square kilometres on one site. In addition to pebbles, there are also vast quantities of clay pot sherds on the soil surface – which may have been for mulch, but which may also have had another purpose.

Clay pot irrigation

The frequent pot sherds within the Anasazi fields suggest that they might have been irrigating with the buried clay pot method. This is an ancient practice, first mentioned by the Chinese author Fan Sheng-chih Shu in the 1st century BC. The method involves placing semi-permeable clay pots in the ground, filling them with water and then covering them with tiles, so that rather than evaporating, the water slowly seeps through the permeable clay pot and into the soil.[71] Farmers in Zimbabwe have built on the Chinese clay pot system by setting up buried clay pipes between the crops, which saves time and effort. Instead of carrying water out into the fields to fill individual clay pots, the farmers can simply pour it into one end of the clay pipe system, and the water is distributed throughout the field. The water slowly seeps through the clay pipes and into the soil, like a combination of ancient clay pots and modern drip irrigation.[72] The large numbers of clay pot sherds found in prehistoric fields in the American southwest suggests that the notion might have arisen independently in America. Clay pot irrigation is a practical way to farm the desert sustainably, making the best of the limited water resources and lessening the problems of salinisation.

The future

There is now almost universal agreement among environmental scientists that the climate is getting hotter. Deserts currently makes up 12% of the earth's land surface, but desert land is expanding in more than 110 countries and the loss of fertile land is threatening the livelihoods and survival of more than a billion people.[73] Water shortage is a crucial issue for the arid regions of Africa, Asia and the Middle East, causing international contention and economic and social divisions between those who have access to sufficient water supplies and those who don't. The issue is becoming ever more pressing as the climate warms and as the desert expands, placing many of the world's poorest people in conditions that are increasingly marginal for their survival.

When Palmer recorded the ruined towns and fields of the Negev, he believed that nothing could resuscitate them. Thus he described the scene:

> Long ages ago, the Word of God had declared that the land of the Canaanites, and the Amalekites, and the Amorites should become a desolate waste; that 'the cities of the Negeb shall be shut up, and none shall open them' (Jeremiah xiii.19) – and here around us we saw the literal fulfilment of the dreadful curse. Wells of solid masonry, fields and gardens compassed round about with goodly walls, every sign of human industry, was there; but only the empty names and stony skeleton of civilisation remained, to tell of what the country once had been. There stood the ancient towns, still called by their ancient names, but not a living thing was to be seen, save when a lizard glided over the crumbling walls, or screech-owls flitted through the lonely streets.[74]

So why did the desert farms go out of use, and when did the Negev become a 'desolate waste'? There are several arguments, and they boil down to 1) climate change, in which desert farming became unviable due to increased aridity (the environmental determinism argument), 2) human mismanagement linked with the 7th century Arab or Muslim Conquest (the anthropogenic argument), and 3) the breakdown in the trade network because of the Muslim Conquest (an economic argument).

The climate change argument was first developed in the early 20th century by Ellsworth Huntingdon and was widely accepted until the 1930s, when it was rejected in favour of the idea that invaders to the region had mismanaged the land.[75] This was a popular notion, and was based on observations that humans have made an unholy mess of a great many environments all across the globe.[76]

The exact timing of the climate changes is important if we are to try and link causes and effects. The evidence tells us that the desert was cooler and wetter in this region from around 2100 BC to AD 500, during the Nabatean and Roman periods.[77] After AD 500, in the Byzantine period, the region became drier, but the desert farm systems reached their greatest extent during this time, despite the fact that the climate had deteriorated.[78]

In the AD 630s the Muslim Conquest put an end to Byzantine control of the Levant, but there is no evidence for the destruction of the cities of the Negev. Avdat was destroyed by an earthquake in the beginning of the 7th century, but was reoccupied a few hundred years later.[79] Nessana (Nitzana) was occupied into the late 7th century, and in Subeita (Shivta) occupation continued into the 8th century, during which time a new church and a new mosque were built.[80] At least 13 mosques were built in the Negev highlands, and although the cities declined, large farmsteads were built in rural areas, including the outskirts of Beer Sheva.

The climate continued to become more arid, reaching a peak around the end of the 1st millennium AD,[81] and by the 10th or 11th century the settlements of the Negev had been abandoned. It would seem that the disappearance of the Negev settlements occurred because of both climate *and* trade. The desert continued to support farming after the Muslim Conquest, but the land was not as intensively populated. This could well be because the Byzantine economic system and trade routes were disrupted. The final abandonment of the desert, however, can probably be ascribed to the aridity of the climate.

So, if the settlements were abandoned because of climate change, does it make any sense to rebuild them? The success of the experimental research suggests that it does, and many Middle Eastern countries are now putting their own desert runoff systems back to work or developing new ones. Qanats are now being restored in Afghanistan, Algeria, Azerbaijan, China, India, Iran, Iraq, Morocco, Oman and Pakistan.[82]

ICARDA has put together a survey of ancient water harvesting systems in West Asia and North Africa, which includes examples from Tunisia, Jordan, Morocco, Syria, Libya, Iraq, Egypt, Yemen and Pakistan[83]. The early systems not only provide water to poor farmers in these regions, but they also give the farmers a degree of self-sufficiency. The government systems sometimes fail to produce enough water, and rainwater harvesting provides a reliable local supply – and villagers have also pointed out that rainwater tastes better. This is not 'going backwards'; this is the intelligent implementation of appropriate technology.

Notes

1 Uzi said this to Jerry and me one night while we were at his house in Eilat, discussing his research in the Timna Valley and the southern Negev.
2 Palmer 1871, 356.
3 Palmer 1871, 352.
4 Palmer 1871, 367.
5 Evenari *et al.* 1971, 95.
6 Evenari *et al.* 1971, 97.
7 Kedar 1957.
8 Evenari *et al.* 1971, 99.
9 Reifenberg 1955.
10 Kedar 1957.
11 Mark Papworth, pers. comm.
12 Lavee *et al.* 1997.
13 Evenari *et al.* 1971, 109.
14 Barker 2000.
15 Bruins and van der Plicht 2005.
16 Avni *et al.* 2013.
17 Evenari *et al.* 1971, 121–125.
18 Evenari *et al.* 1971, 208.
19 Goldsmith and Hildyard 1984.
20 Evenari *et al.* 1971, 194.
21 Evenari *et al.* 1971, 311.
22 Goldsmith and Hildyard 1984.
23 Woolley and Lawrence 2003, 42.
24 Ginsburg 1871; Bennett 1911.
25 Bennett 1911.
26 Wåhlin 1997.
27 Winnett and Reed 1964.
28 Wåhlin 1995.
29 Wåhlin 1995.
30 Fardous *et al.* 2004.
31 Jenssen 2006.
32 For a particularly good summary, see Oweis *et al.* 2004 *Indigenous Water-Harvesting Systems in West Asia and North Africa*. Each chapter describes the water collection and conservation measures of a different country in the region.
33 Karrou and Boutfirass 2004.
34 Rosen 2000.
35 Evenari *et al.* 1971, 337.
36 Evenari *et al.* 1971, 337.
37 Evenari *et al.* 1971, 332.
38 Groves and Hinton 2004.
39 Yair 1983.
40 Evenari *et al.* 1971, 179.
41 Hillel 1991, 117.

42 Hillel 1991, 118.
43 Barker 1996.
44 Clarke criticised over classics. 31 January 2003. http://news.bbc.co.uk/1/hi/education/2712833. stm. Accessed 9 November 2018.
45 Clarke lays into useless history. 9 May 2003. Times Higher Education Supplement. Also reported in The Guardian, https://www.theguardian.com/uk/2003/may/09/highereducation.politics. Accessed 9 November 2018.
46 Santayana 1905, 284.
47 Barker *et al.* 1996, xvii.
48 Gilbertson *et al.* 2000, 139.
49 Gilbertson *et al.* 2000, 154.
50 Gilbertson *et al.* 2000, 141.
51 Hillel 1994, 192.
52 Evenari *et al.* 1971, 178.
53 Wulff 1968.
54 Wulff 1968.
55 Lightfoot 2000.
56 Lightfoot 2000.
57 Wulff 1968.
58 Cressey 1958; Hillel 1994, 192.
59 Pearce 2004.
60 Issar 2014.
61 Hillel 1991, 123.
62 Al-Ghafri *et al.* 2003.
63 Lightfoot 2000.
64 Dutton 1989.
65 Dutton 1989.
66 Hussain *et al.* 2008.
67 Lightfoot 1996.
68 Lightfoot 1996.
69 Rigg and Bruce, 1923; Lightfoot 1994; 1996.
70 Lightfoot 1996.
71 Bainbridge 2001.
72 Gladstone *et al.* Available online at: https://www.doc-developpement-durable.org/file/ programmes-de-sensibilisations/forets-protection/buried_clay_pot_PACE.pdf (accessed 1 August 2017).
73 UNCCD 2009 (Fact Sheet 1).
74 Palmer 1871, 392.
75 Issar and Zohar 2007, 3.
76 Issar and Zohar 2007.
77 Rosen 2000.
78 Rosen 2000.
79 Rosen 2000.
80 Rosen 2000.
81 Rambeau and Black 2011.
82 Manuel, Lightfoot and Fattahi 2017.
83 Oweis *et al.* 2004.

4

Food security

Agricultural landscapes in the Western world have become increasingly industrialised and low in biodiversity, and standard farming practice in developed countries is to grow large fields of single crops sprayed with pesticides and herbicides. Many leading agronomists believe that such industrial methods are the only way we will succeed in feeding a growing global population. However, many opposing agronomists argue that agroecology – or sustainable agriculture – is a more productive and efficient use of land. This chapter reviews the archaeological, environmental and ethnographic evidence for sustainable agricultural land management, as it has been practiced in the past, and as it is practiced today in countries that still use traditional, pre-industrial methods.

A range of evidence demonstrates that small, biodiverse farms are more productive per hectare than agribusinesses that practice monocropping. However, I suggest that there are many potential compromises and small steps that can be taken towards sustainability, and substantial environment benefits accrue where agroecological methods are introduced into industrialised agricultural systems. A key point that I want to make is that traditional varieties of crops and specialised breeds of livestock are important adaptations to marginal environments; it is an essential aspect of our food security that we do not lose these resources, which were developed over thousands of years but which are now vanishing at an alarming rate.

Hunger, poverty and unsustainable agriculture

As the global population rises, there is general agreement that we need more efficient agricultural systems, but there is disagreement about how we should go about creating them. In 2009 The Royal Society published a report on the future of agriculture, in which it was argued that developing new genetically modified products will be essential if we are to feed an expected global population of 9 billion people by 2050.[1] The former American Secretary of Agriculture, Earl Butz, put the argument more bluntly when he stated that, 'Before we go back to organic agriculture in this country, somebody must decide which 50 million Americans we are going to let starve or go hungry.'[2]

Support for industrialised agriculture is based on the assumption that large scale farming is more cost efficient, and yet many experiments have demonstrated that traditional, sustainable agriculture can actually produce *higher* crop yields than industrialised farming.[3] The agricultural systems that produce the highest yields per hectare are not large industrial agribusinesses, but small farms on which different crops are grown together in inter- or multi-cropping systems.[4] Not only do smallholders grow more per hectare, but these higher yields are produced despite the fact that small farms are often located on marginal land with poor soils. The Food and Agriculture Organization of the United Nations (FAO) reports that the most productive farms in a number of countries are in fact the small farms: the most productive in Syria are just half a hectare, in Mexico they are three hectares, in Peru they are six hectares, in India less than one hectare and in Nepal less than two hectares.[5] When farm size increases, production per hectare actually falls.[6]

Smallholders feed a third of the global population, and they provide 80% of the food in the developing world.[7] Such farms are predominantly organic, following traditional, pre-industrial methods such as intercropping (also called polyculture), companion planting, water harvesting, mulching and other such practices that we now regard as forming the basis of the modern science of agroecology. Many farmers are turning away from the industrial techniques introduced through the Green Revolution of the 1960s–1970s, having found that their yields dropped as soil quality diminished, and also having found that the hybrids that were introduced into their ecosystem were not as dependable as traditional varieties. Inorganic fertilisers are expensive and cannot replace organic manures, and crop pests proliferated as beneficial insects were killed along with the crop pests. The good news is that depleted soils and biodiversity are now being restored by a combination of traditional methods together with new research in agroecology, which includes new polyculture combinations and new methods of natural pest control.

I am not suggesting that we reject science – quite the opposite. It is certain that we need to grow more food for a growing population – but we need to define agricultural success in a different way: a successful farm is one that is productive and *sustainable*. It needs to produce abundant crops, even during droughts and other climatic extremes, and it needs to do so without depleting local or global resources. I am advocating a move away from industrialised agricultural practices such as monocropping, over-use of chemical fertilisers and under-use of organic manures. Such practices destroy the soil, pollute our land and water, use vast quantities of non-renewable energy and leave an unacceptably large carbon footprint.

GM crops and hybrids

Experiments with genetically modified (GM) crops have only been running for a few decades, but already there have been a number of problems. GM crops such as maize, soybeans and cotton are bred to be resistant to glyphosate, a herbicide commercially known as Roundup, so that, in principle, fields can be sprayed with herbicide that kills weeds without harming the crop. Unfortunately, over 11 million acres (4,451,542 ha) of farmland in the United States are now infested with glyphosate resistant weeds.[8] One of these 'superweeds', called water hemp, grows up to 8 feet tall, and is poisonous to livestock. Farmers are now having to use more glyphosate than ever, together with a number of other combined herbicides to try and combat the problem. This is costing nearly $1 billion a year, and the problems are only getting worse. According to one study, there are now at least 21 species of arable weeds that are resistant to glyphosate, with an additional one or two species per year developing resistance.[9] Another study cites 68 glyphosate-resistant weeds.[10]

Many farmers are now resorting to weeding the fields by hand, but some are simply giving up. In 2008 it was reported that in the previous year, 10,000 acres (4046 ha) of farmland in Georgia were abandoned because the land was no longer viable due to a glyphosate resistant weed called Palmer amaranth or pigweed.[11] This weed is now thought to be glyphosate resistant on 630,000 acres (25,495 ha) of farmland in Georgia and the Carolinas.[12] The USGS reports that 11,000 tons of glyphosate were used in 1992, but by 2007 that number had increased to 88,000 tons.[13] This report goes on to note that in areas where it is used most intensively, glyphosate is now found in the air, the rivers, and even in the *rain*. While Monsanto argues that it is a harmless substance, reports of birth defects and cancer have led to studies which indicate that it may actually be quite dangerous.[14] The World Health Organization expressed concerns, but later concluded that glyphosate is 'unlikely to pose a carcinogenic risk to humans from exposure through the diet'. The lack of consensus on GMO safety was recently flagged up by the European Network of Scientists for Social and Environmental Responsibility, and it is to be hoped that further research will clarify the issues.[15]

One of the problems is of course that scientific research can tell us probabilities but rarely provides us with absolutes; it is possible to demonstrate that a substance is harmful, if many researchers are able to replicate the initial study that found it to be so, but it is difficult to prove that a substance is safe. The scare over vaccines was a case in point: one researcher provided one study that seemed to show a link between the Measles, Mumps and Rubella inoculation (the MMR jab) and autism, but nobody was able to replicate the study. People became anxious and thought the authorities were lying to them or covering up this one flawed and now discredited study, when in fact the authorities were trying to explain to the public that a) there was copious evidence for the safety of the MMR, b) that science works in probabilities, not certainties, so scientists can't 'just tell people the facts', and c) that if a study can't be replicated, it means that it cannot be regarded as convincing.

Whether GM crops and Roundup are safe or otherwise, I would argue that it is a worry that GM traits are now spreading into non-GM plant species. In Scotland, for example, GM potatoes were bred that carry snowdrop genes, which were added to make them resistant to potato aphids. The unintended result was that ladybirds had their lifespan reduced by half.[16] Since each ladybird eats *c.* 1000 aphids in its life (as well as mites, scales, whitefly, mealybugs, cabbage moths, bollworms, tomato hornworms and broccoli worms), this is not a happy outcome.

Genetically modified cotton (Bt transgenic cotton) is created using toxins from a bacteria called *Bacillus thuringiensis*, which occurs naturally in the soil. It was introduced into China in the late 1990s, and initially it appeared to be a success. The new strain was created to resist infestation by bollworm larvae, and there was a 60% drop in pesticide use in the early years.[17] Unfortunately, the bollworms that attack cotton plants are now developing resistance to the Bt toxin, and there is a rise in secondary pests, and there has also been a huge decrease in the numbers of insects that prey on the bollworm.[18] There is now a plague of mirids attacking the cotton on the GM cotton fields of 5 million Chinese farmers, who are fighting back by using vast quantities of pesticides.[19]

In parts of India, Bt cotton has been less successful than traditional strains.[20] Not only did it fail to protect the plants from the bollworm, but its use led to a 250–300% *increase* in attacks by a range of other crop pests, and it was not resistant to drought, and it returned lower yields than the traditional strains.[21] There is fierce debate about the various field trials, with both sides arguing that statistics are on their side. Proponents of GM technology argue that those against it are anti-science and are ignoring the evidence, however even Monsanto has now admitted that the pink bollworm in Gujarat has developed resistance to Bt cotton.[22] In America, Bt cotton has also suffered attack from the bollworm it was supposed to prevent.[23] Since this report is published in *Science*, one of the most rigorously peer reviewed journals in the world, this evidence should not be lightly dismissed.

The companies that create GM crops and the politicians who support them argue that people in developing countries need GM crops, and that we must provide them with these products or else they will die. There are mixed feelings about GM crops in many developing countries, including a great deal of strong resistance – particularly to the concept that living things can and should be patented. Dr. Tewolde-Berhan, the General Manager of the Environmental Protection Authority in Ethiopia, has stated '...we strongly resent the abuse of our poverty to sway the interests of the European public'.[24]

The concept of copyrighting plant varieties was agreed by the World Trade Organization as part of its Trade Related aspects of Intellectual Property Rights (TRIPS) agreement. This was challenged by a collective of developing countries, largely on the basis of the right to medicine. If traditional plants are investigated by biotechnology companies, and the 'active ingredient' genes are identified and copyrighted, then the plants have in essence become 'privatised'. Large corporations make money out of the plants developed by indigenous people, and the people whose ancestors developed the plants are no longer regarded as having ownership. Medicine is only one aspect of the issue, and Tewolde-Berhan has argued that patenting living things will contribute to a decline in biodiversity, which is obviously in nobody's best interest, certainly not small farmers.

Another issue is the necessity of buying seeds every year, instead of gathering them from each year's harvest and sowing them the following year. The GM companies (mainly Monsanto, which owns over 90% of the world's GM crop varieties) have patents on their seeds, which means that it is illegal to collect seeds directly from the GM plants a farmer has grown – but hybrids don't produce dependable seeds anyway, so this is more of a Green Revolution issue than a GM issue. Nevertheless, buying seeds each year is a major expense to farmers in developing countries, and it stops farmers from experimenting and breeding their own varieties, as they have traditionally done. Peasant farmers are bankrupting themselves buying hybrid and GM seeds, and the expensive pesticides, fertilisers and herbicides that must be used with these specialised crops. One of the proponents of Bt cotton argues that if it were not a success, Indian farmers would not continue to buy it,[25] but a farmer in Maharashtra sums up the situation by pointing out that, 'We cannot afford planting Bt anymore, but we cannot buy non-Bt seeds in the market. The dealers tell us that there is no supply.'[26] It has been suggested that agrarian bankruptcies have led to a wave of suicides, with the deaths of more than 20,000 Indian farmers in the last few years.[27] This is a humanitarian crisis of horrendous proportions, and although there are also those who argue that the suicides have nothing to do with GM crops,[28] it is clear that there are problems with the Indian agricultural economy. The price of the seeds, pesticides and fertilisers are a major issue. Water is another.

A problem introduced by the Green Revolution was that the new hybrid crops use five to ten times more water than traditional varieties,[29] and Bt cotton requires even more.[30] The traditional water harvesting systems in India formerly enabled villagers to store water and to maintain a high water table for wells, livestock and crop irrigation, but these localised systems were disrupted by the introduction of deep wells which depleted the aquifers. In villages where the government has not yet brought in the centralised 'improvements' to water supply, villagers using traditional methods have had plenty of water during recent droughts that left their more developed neighbouring villages in need of emergency tanker trucks.[31] A similar discovery was made in an experiment to reconstruct the Byzantine or earlier water harvesting systems in the Negev desert in the early 1960s. The Negev experiment was set up just before the most extreme drought ever recorded in the region, but while industrialised crops withered and died, the farms dependent on water harvesting and storage succeeded in producing a harvest.[32] Traditional water harvesting systems are now being restored in many countries,[33] India included, and given the crisis in the Indian countryside, this cannot happen quickly enough.

India is not passively accepting the arrival of GM crops. The former cabinet minister for rural development, Dr. R. P. Singh, has argued that allowing GM crops into the country is 'anti-farmer and anti-people'. He makes the point that India needs solutions that are 'economically, socially and environmentally sustainable' – not GM crops, which require expensive inputs and which may cause ecological damage that would be irreversible. India is currently making great leaps towards sustainability, drawing on a long history of sophisticated sustainable agriculture, crop breeding and water harvesting. These systems were disrupted by the Green Revolution, which aimed to improve yields and to modernise agriculture in India, but villagers found that the new hybrid crops caused allergies, required expensive fertilisers, and were not resistant to drought.[34] Furthermore, the hybrid sorghum that was introduced exhausted the soil, and inorganic fertilisers could not replace lost soil organic matter.[35]

In Mexico, the peasant farmers are also feeling that they are under attack by agribusiness. Maize probably has its origin in a grass called teosinte, which grows naturally in Mexico and Guatemala. Analyses of phytoliths (silicate structures in plants) and starch grains have demonstrated that maize occurs in the archaeological record by 6700 BC, first appearing in the Balsas River Valley in Mexico.[36] Over the course of thousands of years, maize has been further developed in order to enhance certain characteristics, such as size and colour. The early farmers thus developed hundreds of varieties of maize, which they regard as their most important cultural heritage (Figure 4.1). These varieties are now being contaminated by GM maize, largely because Mexico imports maize from the United States, where the prices are kept artificially low.[37] A further problem is that in the American exports, GM and non-GM

Figure 4.1: Maize varieties. (Photo © Roger Costa Morera/Shutterstock)

maize are not separated.[38] Also, some Mexican farmers have planted GM maize which was given to them as food aid.

Aldo González, a spokesman for the Union of Organizations of the Sierra Juarez Oaxaca, has stated that:

> Native seeds are a very important part of our culture. The pyramids may have been destroyed, but a handful of maize seed is the legacy we can leave to our children and grandchildren. Today they are denying us this possibility. The process of globalisation that our country is going through and the undermining of governmental authority are keeping indigenous communities from being able to pass on this age-old legacy, which represents more than 10,000 years of culture. For 10,000 years our seeds have proven they don't harm anyone. Today they're telling us that transgenic seeds are harmless. What proof do they have of this?[39]

Since the Mexican government is unprepared to ban GM maize imports, in 2011 two of the Mexican states that are major maize producers – Tlaxcala and Michoacán – have independently banned GM maize in their regions.[40]

Traditional or 'folk' varieties: safety in biodiversity

Modern wheat species have been bred to be extremely productive, but for much of our history plants have been bred for other characteristics, chiefly ones that make them more reliable. The modern practice of monocropping, or planting large fields with a single species, would have left prehistoric farmers extremely vulnerable to crop pests, disease and drought. It is far safer to plant different, related crop species together in a single field – a practice known in historical times as 'maslin'. In the Medieval period wheat and rye were often grown together, rye being more resistant to frost, or alternatively maslins were grown of wheat and barley or barley and oats.[41] The practice has been observed in many different present-day ethnographic studies, for example in Greece.[42] On the Greek island of Amorgos, farmers grow up to four types of wheat and two types of barley together.[43] In Ethiopia, renowned for its many different strains of drought tolerant wheat, up to fifteen varieties of wheat are grown in one field.[44]

Maslin agriculture is based on the premise that different species are resistant to different climatic extremes, and there is safety in creating biodiversity within a single crop. Whereas some varieties are resistant to drought, for instance, others are resistant to particular diseases, pests or fungi. In Greece, farmers buffer against drought by growing a maslin of bread wheat, which is better at withstanding frost, with durum wheat, which is more resistant to drought.[45] In Ethiopia, there are a vast number of wheat varieties that are resistant to heat and drought,[46] and in Georgia the low yielding but cold tolerant wheat variety *Triticum carthlicum* is grown in a maslin with varieties that are resistant to fungal attack.[47] The resilience of folk varieties, the dependable nature of the yield and the resistance to local conditions are all factors that encourage farmers to grow locally bred crops. In the Aras Valley in Turkey, for instance, farmers report that although *Triticum carthlicum* provides a meagre harvest, other varieties won't grow at all in their cold-affected region.[48] Turkish farmers also grow wheat together with rye, because rye is better adapted to the frost that makes mountain farming so difficult.[49]

The practice of growing multiple species or varieties together is also evident in the archaeological record. Charred cereal remains from the Neolithic onwards are often mixed assemblages, and it may be that this is actually the case for *most* prehistoric archaeobotanical assemblages.[50] The recovery of a mix of cereal varieties in one archaeological deposit is not proof that they were necessarily grown together, but one study has convincingly demonstrated that emmer and spelt (two early wheat varieties) were grown together in the British Iron Age; there was a strong correlation between the crop species and the accompanying weed seeds, which showed that the different wheat varieties were grown in the same field.[51] A study of a late Bronze Age settlement at Stillfried, Austria found four different types of wheat in one pit, with traces of a fifth,[52] and at the Neolithic Turkish site of Can Hasan, a 'huge variety' of different strains of wheat were grown.[53]

In Southeast Asia, farmers traditionally planted dozens of different varieties of rice, and although many now grow large fields with single rice varieties for export, they still grow the traditional varieties for consumption at home.[54] In North Thailand, farmers grow rice varieties that ripen at different times, so that if a heavy storm flattens one crop, the others still stand a chance.[55] A study of Mende farmers in the 1980s in Sierra Leone reported that they grew 70 different varieties of rice, while Susu farmers in Sierra Leone mix African and Asian rice varieties.[56] In Malawi, small farmers say that they plant a wide variety of beans to ensure that they always have a yield. Peruvian farmers grow a wide range of potato crops, and as altitude increases they use a correspondingly higher proportion of traditional varieties, because they are better adapted to the conditions than modern strains.[57] Hopi farmers in the American southwest still grow many different types of maize, with one study finding 21 varieties on a single farm, including seventeen traditional ones.[58] The Hopi blue maize, for instance, can be planted deeper than most varieties – up to 25 cm deep. This adaptation allows it to grow in the desert, where water is a limiting factor. In India, there are a number of wheat varieties that are resistant to salinity, and so grow well in areas that are occasionally flooded by the sea.[59] Unfortunately, many traditional strains were lost during the Green Revolution.

In optimal growing seasons traditional varieties are not always as productive as modern strains, but in the environment where they were created the traditional varieties often produce higher average yields than modern crop species.[60] In bad years, they are far more likely to produce a yield than modern varieties, and some traditional varieties have been found to outperform modern crops in every respect; research on rice in the Philippines found that folk varieties produced higher yields than modern ones, and similarly in Ethiopia the traditional outperforms the modern.[61] After Hurricane Mitch hit Central America, the farmers who were practicing sustainable methods suffered far less than conventional monocrop farms. The sustainable farms were buffered against severe storms by practices such as growing cover crops, intercropping and agroforestry, which meant that they had 20–40% more topsoil.[62]

Traditional breeds of animals are also being lost at an alarming rate. It has been estimated that 700 breeds of livestock became extinct in Europe in the 20th century alone.[63] We are losing six breeds of livestock per month, and 30% of the breeds developed over thousands of years are in danger of extinction.[64] This loss is as dangerous as the loss of traditional crop varieties, and for the same reasons. Animals have been bred to withstand a range of conditions, including subsistence on poor quality feed and survival in extreme aridity, and replacing these breeds with higher producing but less resilient animals is a very risky strategy for subsistence farmers. Livestock will need to withstand a range of diseases and environmental extremes, especially as the climate becomes even more arid in the desert regions.

Intercropping

Planting multiple varieties of wheat or rice is one way of maximising security, but planting completely different types of crops together – known as intercropping or polyculture – can increase productivity exponentially (Figure 4.2). A classic example is what the American Indians called the Three Sisters: maize, beans and squash. The maize grows first, and provides a beanpole for the beans to grow on. The beans, like

Figure 4.2: Garden polycropping in California. (Photo © Ros Creasy)

all legumes, are nitrogen fixers, so they provide the key nutrient for the maize. The squashes have large leaves that very effectively cover the ground, reducing soil erosion by protecting the soil from rainsplash (the destructive impact of raindrops, which break up the soil structure). The squash leaves also provide shade, which reduces evaporation from the soil, and the shade also represses the weeds. Furthermore, squash leaves contain toxic chemicals that are washed into the soil by rainfall; these chemicals prevent weed growth but do not harm the maize and beans. The maize-beans-squash intercropping system sustains large numbers of a type of wasp that eats insect pests, so that leaf hoppers, caterpillars and thrips do not cause as much damage as they do in monocropping systems.[65]

In Mexico, where the Three Sisters system is commonly used on smallholdings, it produces 73% more food per hectare than farms growing maize alone.[66] In modern parlance this mixing of complementary plants is known as companion planting, and we know from the early documentary sources that the Three Sisters system was widely practiced in North America. At least half of the early accounts dating to the days of the first contact of Europeans and North American Indians mention intercropping, usually involving the Three Sisters.[67]

Intercropping is not an outdated practice out of the distant past. People still practice this traditional form of agriculture all over the world today, and it is very successful even in modern terms. In Southeast Asia, legumes such as peanuts are typically grown in rotation with rice. The legumes provide nitrogen for the rice, and are grown after the rice is harvested so that they can take up the 'leftover' water from the paddy fields. In the Gambia (West Africa) farmers plant peanuts together with sorghum, because the weevil that usually attacks the sorghum will focus on the peanuts instead; in fact, sorghum cannot be grown at all in this area if it is not intercropped with peanuts.[68] Legumes are commonly grown together with grains; 98% of the cowpeas grown in Africa are intercropped with other plants, and 90% of the beans grown in Columbia are intercropped. In experiments of multi-cropping with sorghum, peanut and millet combinations, yields were higher than the same species mono-cropped, and when subject to controlled drought conditions they outperformed mono-crops to an even greater degree.[69] All of these systems are geared to providing food security.

There is a group of indigenous people in Mexico called the Wixaritari, who believe that the world functions as a magic circle in which nothing can operate alone. We share this belief in the West, but we call this magic circle 'ecology', a subject so complicated that when its academic founder, Eugene Odum, first proposed to teach it at University he was actually laughed out of the room by his colleagues. Now, we recognise that all things in nature are interdependent to the point where you cannot alter one part of an ecosystem without having an effect on another. Sudden, drastic changes to the ecology can cause innumerable problems, for instance the widely known example of cane toads that were introduced to Australia. The idea was that they would eat the beetles that attack sugar cane crops, but not only did they fail to control the beetle pests, they then made matters worse by turning on other species. The toads have no natural predators, and they contain toxins that can kill most of the native animals that are unfortunate enough to try eating them.

Agriculture is an area in which non-native species are regularly introduced into new ecosystems, and the results depend on how well the plants fit in to their new neighbourhood. In sustainable agriculture, or agroecology, the aim is to grow plants as part of a whole ecosystem, not as a single crop on its own in a field. Sustainable agricultural systems attempt to mimic nature in that each species has its own ecological niche. Plants are not competing with each other, but benefitting each other by providing shade, fixing nitrogen and attracting pests that would otherwise attack their neighbours. In the tropics, such systems can be extremely complex. Plants are carefully selected for their micro-environment, so shade tolerant species are planted on the forested edges of a plot, while plants that need more moisture are placed at the bottom of slopes. Plants that need a lot of nutrients are planted on old ash dumps, and climbing species are put next to trees or other upright plants that can act as supports.[70]

The beauty of the intercropping systems is that different crops take up different nutrients, and in addition, the leguminous plants are providing nutrients for their neighbours.[71] If this sounds friendly, it is worth noting that they fix nitrogen with the help of bacteria that live on their roots. In biology, this sort of mutual beneficence is known as 'symbiosis' or (more recently) 'symbiotic mutualism', and it has been argued that there is far more symbiotic mutualism in nature than we realise. At the risk of sounding romantic, nature is not all 'red in tooth and claw'.

Trees are an underused resource that can be intercropped very successfully with other crops. A lot of erosion takes place on hilly ground, and planting tree crops for fruit and nuts is a good use of steep slopes or dry and rocky ground that is unsuitable for agriculture,[72] but trees have other benefits as well. Bringing trees into the agricultural system in Africa has actually been shown to increase maize yields; the leaf litter provides vital organic matter, and the trees also protect the crops from wind.[73] One particular tree, the Faidherbia, is a nitrogen-fixing leguminous tree that is native to Africa and the Middle East. It functions as other legumes do, by taking nitrogen out of the atmosphere and making it available to other plants in the soil, but it also produces leguminous pods, which make excellent animal fodder, while the seeds themselves are also eaten by people.[74] Faidherbia flowers appear several months after the beginning of the dry season, and so provide pollen for bees at a time when other sources are declining. It also contains fluorine in the bark, which is used to clean and strengthen teeth, and it coppices well, and can be used for firewood. It loses its leaves in the rainy season, and so it does not compete with crops for sunlight.

Some of the most elaborate intercropping systems were developed in China. Agricultural treatises written in the Warring States period (475–221 BC) describe crop rotation, flood control, irrigation and manuring processes that took place over 2000 years ago.[75] An agricultural treatise written in the 6th century AD describes crop rotation and the benefits of planting legumes in advance of other crops,[76] which as we know adds nitrogen to the soil. A particularly efficient system was developed in the Guangdong Province, on the Zhujiang (Pearl River) Delta,[77] beginning about 600 years ago.[78] The rice paddies double as fish ponds for grass carp, which eat only the weeds that grow in the paddies.[79] Common carp and dace are kept in the paddies to eat the grass carp excrement, which would destroy the water quality, and the other fish waste settles to the bottom of the paddy[80]. Periodically, the nitrogen-rich mud at the base of the paddy is collected and used to build fertile raised fields around the paddies, where mulberry trees are grown. The mulberry leaves are used to feed silkworms, the bark is used to make paper, and silkworm waste and mulberry leaves drop into the ponds and provide food for the fish.[81] After the last crop is in, the mud from the paddy is used to fertilise the surrounding dyke, where they grow more than 40 varieties of vegetables and a variety of fruit trees.[82] In some of the southern provinces of China, up to ten vegetable crops are grown each year on the same land.

The Chinese still practice integrated farming with animals, including the system of grass carp in the rice paddies, described above, but also including other animals. In some systems, different types of fish are raised, for instance grass carp excrement provides food for plankton, and plankton provide food for silver carp.[83] Ducks raised for the market are also useful for controlling weeds in the rice paddies.

Animals in these systems can also provide useful pest control. In Java, ducks are kept in the rice paddies to eat the insect pests, and chickens are kept in the gardens to eat the insects, particularly the leaf-rolling and leaf-eating pests of peanut plants.[84] The caterpillars that infest their gardens are captured and used for fishing bait, or sold on the markets as food for chickens, pond fish and caged birds.

Animal pests that eat the crops can also be regarded as food. In Indonesia they eat the grasshoppers that they catch in the rice fields, or sell them as bird food; they also eat the birds that attacks their rice paddies, as well as squirrels and termites that damage their crops.[85] In northeast Thailand the farmers go out at night with torches which attract the beetles that eat their sesame crop; the beetles are captured and eaten or sold as food. In Borneo the crops are subject to severe damage by wild pigs, but these are killed and eaten. In medieval Britain rabbits were a rather frail species that had to be nurtured in managed warrens, but over time they became much hardier.[86] By the 18th century they had become a crop pest, but a very edible one; by the 19th century, they had become an important food source for the poor.[87]

The Chinese traditions predating the Cultural Revolution are characterised by careful use and reuse of all resources. The Cultural Revolution caused a lot of ecological damage, *e.g.* deforestation and soil erosion, but there is now a renewed interest in sustainable agriculture – and the new, freer economy now allows farmers to sell their surplus crops after they have sold the requisite amount to the government. This means that there is a financial incentive for farmers to maintain the extremely productive intercropping systems of the past, because these systems allow farmers to produce vast amounts of food on very small parcels of land. In fact, in the 1980s

China was feeding over a billion people – which is 22% of the world's population – on just 7% of the world's arable land.[88] They do this by maintaining the old traditions of intercropping, but there are some interesting changes, for instance, in the 1980s it was discovered that draining the rice paddies halfway through the growing season causes the paddies to emit significantly less methane (a greenhouse gas).[89]

One of the interesting things about the early Chinese agricultural treatises is the emphasis on balance and sustainability; on working *with* nature instead of against it. The interaction of plants with one another (and with animals) is of great importance, and continues to be a significant element of organic farming today. Planting different strains of rice in the same field can reduce the spread of disease, and is a long-standing tradition in Asian countries.[90] It also increases the likelihood of crop survival during environmental extremes, which some strains are able to withstand better than others. In diversity is security.

Reintroducing traditional agriculture

In the early 20th century, the Japanese government led a campaign *against* rice crop diversity. The aim was to modernise, and the result was the loss of 800 strains of rice and a reduction from 1200 different varieties grown in 1910 to around 400 in 1920. The Green Revolution of the 1960s and 70s accelerated this global trend, and by 1983 more than half of the wheat and rice crops grown in developing countries were modern varieties. By 1986 over half the maize in developing countries had gone the same way. In developed countries, the industrialising trend continued. In America, small farmers could not compete with agribusiness, and vast areas of land were turned over to monocropping. In the UK, hedgerows were torn out with government support, so that field size could be increased and farming could be mechanised.

There is now an increasing interest in reversing this trend, and as the global climate warms and desertification spreads, arid regions are a matter of particular concern. In Senegal, West Africa, The Rodale Institute has helped introduce sustainable systems that are proving to be both productive and resilient. Senegal lies within the Sahel, a band of semi-arid grassland where the soils are sandy and desertification is a major problem. Adding inorganic fertiliser and pesticide is too expensive for most of the local farmers, but even if these inputs were more affordable, they are not appropriate in this environment. There is so little organic matter in the sandy soil that there is nothing to bind the nutrients, and the fertiliser simply washes away in the rain or is taken up by the weeds and soil microbes. Local farmers have started integrating livestock into their farms, so they can improve the soil with manures, and they have begun growing legumes, which add nitrogen to the soil. Composting and water harvesting have also helped, and together these sustainable improvements have pushed up yields by 75–195%.[91]

The desertification of the Indian countryside has also been turned around by projects in a number of regions, including the Deccan plateau. The climate in the northern part of the plateau is semi-arid, and low rainfall and a lack of organic material has reduced much of the farmland to dust. With the support of the Deccan Development Society, this is now being reversed by a collection of women who have formed voluntary village associations called Sanghams. Most of the women in these groups are from the lowest caste, the dalit or untouchables, and before they got together they lived in poverty and isolation.

The Deccan has a tradition of growing drought-tolerant strains of barley, wheat, sorghum and millet; this system dates back to the Chalcolithic (*c.* 2200–1800 BC) or perhaps even earlier, but traditional varieties were abandoned in favour of hybrids during the Green Revolution.[92] Now, the 5000 women in the Sanghams have brought back 80 traditional crop varieties, most of which had ceased to be grown in the area. By planting different species together (sorghum, pigeon peas, pulses, amaranth, fibre crops and cattle feed), they are now producing exponentially more food (*e.g.* up to 22 species on one three acre plot) (Figure 4.3). Self-mulching plants control the weeds, which reduces the labour required. The cattle provide manure to improve the soil, and earth banks and rock dams help retain water. This keeps the ground water levels higher, and villagers downstream from these fields are finding they have more water in their wells. The women have rejuvenated over 10,000 acres (4046 ha) of degraded

Figure 4.3: Polycropping on the Deccan Plateau, India. (Photo courtesy of the © Deccan Development Society)

land, and are now raising over three million kilos of grain each year – six times more than under Green Revolution methods.[93] They are running community seed banks with a huge range of varieties, so that seeds are available for everybody – not just the wealthy. They have planted medicinal herb gardens for whole communities, and they have planted over a million trees.[94] This is a system in which everybody benefits. Most of the methods are traditional, and only went out of use around 40 years ago, but low caste women now have enough food and medicine, and also some financial autonomy. Their children can now go to school, and escape from the bonded labour and illiteracy that used to hold back the dalit class.

Here is another remarkable combination of old and new ideas. In Bolivia, small farmers in the high altitudes were struggling with the erosion of sandy soils on steep slopes. After experimenting with traditional farming methods, they hit on the practice of using the local lupin (*Lupinus mutabilis*) as a green manure.[95] Lupins are a legume, so they are nitrogen fixers – but they fix 25% more nitrogen than clover and 28% more than other types of peas and beans, so they are actually much more efficient nitrogen producers than the crops that are usually sown as green manures. They also grow at high altitudes, where the cold temperatures discourage the usual nitrogen fixing crops. *Lupinus mutabilis* grows naturally in the Andes, so it does not cause problems by becoming invasive, and the protein-rich seeds have been eaten by the Andean people for millennia.[96] Bolivian farmers are growing lupins to improve their soils, and have seen their potato crop yields increase by 378% using lupin green manure alone. When they also added sheep manure, their yields increased from a pre-lupin rate of 1780 kg per hectare to 13,000 kg/ha, which is a 630% increase.[97] Elsewhere in Bolivia, the introduction of organic farming, the reintroduction of local potato varieties and crop rotation with nitrogen fixers has increased production by 150–375%.[98]

These are just a few examples among hundreds. In Mexico, 100,000 small farmers have gone organic, and have experienced a 50% increase in their yields.[99] This trend is not restricted to developing countries. Before 1989, Cuba had the most industrialised agricultural system in Latin America. The collapse of the Soviet Union meant that they lost their key trading partner, which left them short of food and short of money for fuel, pesticides, herbicides and fertiliser.[100] The small farmers were able to boost production very rapidly by returning to pre-industrial methods, but the State-owned industrial farms struggled. By turning the large farms over to small farmers' co-operatives, production was rapidly increased by a return to intercropping, together with new methods of biological pest control.[101] By 1996–97, Cuba had the highest food production on record for the 13 staple foods in their diet.

The surge in productivity in Cuba did not take place overnight. The country was catapulted into sustainable agriculture quite suddenly, and initially there were shortages that resulted in malnutrition following the economic collapse. It typically takes three to eight years to go from industrialised to organic farming, and Cuba had to make the transition quite abruptly. Fortunately, Cuba had already begun a program of research into sustainable agriculture, which they implemented very rapidly.

The Cuban Ministry of Agriculture developed a host of different approaches, and took the sensible step of seeking advice from older farmers. The new, sustainable methods that they introduced (and reintroduced) included polycropping, intercropping and planting catch crops to lure pests away from the food crops. Banana stems were painted with honey to attract insects away from the crops – a method that proved to be very effective in fighting the sweet potato weevil.[102] The Cubans also planted pest-tolerant varieties of crops, and they used the natural pesticides produced by Neem trees.

Natural pesticides include bacterial and fungal insect diseases that kill pests but do not harm people; an array of useful bacteria and fungus species were mass produced in the new laboratories that sprang up around the country. Parasitoids also proved to be useful: these are insects whose larvae live as parasites that eventually kill their hosts, which in this instance were crop pests. Predatory wasps and ants were grown in laboratories and distributed to farmers, the ants helping to control sweet potato weevils – although it was also discovered that growing sweet potato together with maize can eliminate the need for pesticides altogether.[103]

The use of bacteria in producing natural fertiliser is one of the more remarkable developments in Cuba. Scientists are mass producing bacteria that stabilise phosphorus in the soil – something never done before on a mass scale.[104] They are also producing nitrogen fixing bacteria, including *Rhizobium* species, which are bacteria that live on the roots of plants, working in beneficial symbiosis by taking nitrogen out of the atmosphere and converting it to ammonium, which is a form that is available to plants. *Rhizobium* bacteria are grown in labs, so they can be added to arable soils, where they will hook up with crop plants and thus enable them to create their own nitrogen. There is now a further addition to this industry: mass production of *Azotobacter*, an extraordinary bacteria that takes nitrogen out of the air and deposits it in the soil, so it is available to plants – but this bacteria lives independently in the soil, and is not dependent on host plants such as legumes. In 1989 Cuba was using more nitrogen fertiliser per hectare than America, and now, Cuban scientists have found a way to help soil create its own nitrogen.

Natural pest management in America is lagging behind Cuba, but interest is growing. One method of biological pest control that is practiced widely in California is the use of predatory mites, which are used on up to 70% of California strawberry fields; these mites are very effective in suppressing the spider mites that otherwise damage the crops.[105] Polycropping is another way to limit the damage done by insect pests, and in America it is now being done on an industrial scale (Figure 4.4). Crop rotations also keep down pests and crop diseases, and they are a method that is easy to employ on an industrial scale. Farmers in Michigan have found that when rotating soybeans, wheat and then maize, the soybeans add nitrogen to the soil, which benefits the subsequent wheat, which grows densely and suppresses the weeds in advance of the maize crop.[106] Ohio farmers have found that growing wheat in rotation with

Figure 4.4: Sustainable agriculture in the American Midwest. The contours collect water and prevent soil erosion, and the maize is intercropped with alfalfa, which provides fodder for the cows. (Photo © Jim Richardson).

soybeans also keeps down the potato leaf hopper pest. Crop rotations are relatively easy adaptations that can be carried out on large, industrialised farms.

Irrigation practices could be improved in America, where quite a lot of water is wasted. Switching to a system of drip irrigation would be a good start; this method delivers small quantities of water directly to the plant at the soil surface, thus saving water, limiting weed growth, reducing the likelihood of waterlogging and salinisation and putting less stress on crops.[107] The drought in the summer of 2012 was devastating to crops in the United States and Canada, but farmers who take an interest in sustainable agriculture have developed a range of adaptations to help them cope with the increasing aridity. More drought-tolerant species have been introduced on some farms, including the natural prairie grasses that can be fed to livestock, and more livestock have been introduced to take advantage of the increase in fodder production.[108] Some farmers are staggering the time of planting to increase the likelihood of some crops surviving, and planting perennial crops improves soil quality and moisture retention.[109]

We have a lot to learn from the traditional agriculture of what we call 'developing countries', which have a lot to teach us about adaptation to environmental stresses.

New research in sustainable agriculture can be used to supplement these established methods, and fortunately not all traditional crop varieties have been lost: the US Department of Agriculture has a store of 60,000 varieties of wheat, barley, oats and rye,[110] so we can reintroduce these traditional varieties or use them to create new strains. I am not advocating a wholesale return to past technologies or a return to growing only traditional plant and animal varieties, nor am I saying that everything was better in the past. What I am suggesting is that we can use some aspects of early technology, together with new discoveries in agroecology, to create a healthier, more sustainable and environmentally richer planet.

Notes

1   Royal Society 2009.
2   Lockeretz 2007.
3   Erickson 1988; Parrott and Marsden 2002; Pretty 1998a, 84–90; 1998b; Pearce 2001; 2002.
4   Altieri 2002; Francis 1986; FAO 1980.
5   FAO 1980.
6   FAO 1980; Goldsmith 2003.
7   IFAD 2012.
8   Gillam 2011.
9   Gillam 2011.
10 Benbrook 2009.
11 Robinson 2008; Benbrook 2009.
12 Robinson 2008.
13 Capel and Capelli 2011; Chang *et al.* 2011.
14 Benachour and Séralini 2009; Gammon 2009; Paganelli *et al.* 2010.
15 ENSSER 2013.
16 Pretty 1998a, 5.
17 Nakanishi 2004.
18 Xinhau News Agency 2002; Zhao *et al.* 2011.
19 Coghlan 2006.
20 Sahai and Rahman 2003.
21 Shiva 2004.
22 Monsanto 2010.
23 Kaiser 1996.
24 Masood 2001.
25 Herring 2009.
26 Byatnal 2012.
27 Shiva 2004.
28 Herring 2009.
29 Shiva 2004.
30 Herring 2009.
31 Goldsmith 2003.
32 Evenari *et al.* 1971, 191; Guttmann-Bond 2010.
33 Oweis *et al.* 2004.
34 Henderson 2001.
35 Henderson 2001.
36 Piperno *et al.* 2009; Hastorf 2009.

37 Quist and Chapela 2001.
38 Ribiero 2004.
39 Ribeiro 2004.
40 Latin America Press 2011.
41 Slicher van Bath 1963; van der Veen 1995.
42 Halstead 1990; van der Veen 1995.
43 Jones 1990.
44 Tesemma 1991.
45 Halstead 1990; van der Veen 1995.
46 Tesemma 1991.
47 McLaren 2000.
48 McLaren 2000.
49 Behre 1992, 149; van der Veen 1995.
50 van Zeist 1968.
51 van der Veen 1995.
52 Kohler-Schneider 2002.
53 McLaren 2000.
54 Capistrano and Marten 1986.
55 Capistrano and Marten 1986.
56 Cleveland, Soleri and Smith 1994.
57 Brush 1980.
58 Cleveland *et al.* 1994.
59 Deb 2009.
60 Cleveland, Soleri and Smith 1994.
61 Cleveland, Soleri and Smith 1994.
62 Altieri 2002.
63 Pretty 1998a, 33.
64 FAO 2004.
65 Scialabba and Hattam 2002.
66 Pearce 2001.
67 Doolittle 2000, 141.
68 Vandermeer 1989.
69 Altieri 2002.
70 Norman, Pearson and Searle 1984.
71 Vandermeer 1989.
72 Moinar *et al.* 2013.
73 Altieri *et al.* 2012.
74 World Agroforestry Centre: Agroforestree Database. Available online at: http://www.
    worldagroforestry.org/treedb2/speciesprofile.php?Spid=1 Accessed 9 November 2018.
75 Wittwer *et al.* 1987.
76 Wenhua 1993.
77 Luo and Han 1990.
78 Ruddle and Zhong 1988.
79 Luo and Han 1990.
80 Luo and Han 1990.
81 Luo and Han 1990; Ruddle and Zhong 1988.
82 Ruddle and Zhong 1988.
83 Ruddle and Zhong 1988.
84 Brown and Marten 1986.
85 Brown and Marten 1986.

86 Rackham 1986, 47.
87 Rackham 1986, 48.
88 Wittwer *et al.* 1987.
89 https://www.edie.net/news/0/Chinese-rice-farmers-reduce-methane-emissions-from-paddies-by-40/6434/ (accessed 19 July 2017).
90 Capistrano and Marten 1986.
91 Scialabba and Hattam 2002, 155.
92 Schug 2011, 7.
93 Deccan Development Society. Available online at: http://ddsindia.com/ Accessed 9 November 2018.
94 Henderson 2001.
95 A green manure is a crop that protects the soil from erosion in between food crops. Before sowing the crop, the green manure crop is ploughed into the soil to add organic matter to the soil.
96 Eastwood and Hughes 2008.
97 Scialabba and Hattam 2002.
98 Ruddell 1995.
99 Pretty 1998b.
100 Parrott and Marsden 2002; Rossett 2000.
101 Rossett 2000.
102 Viljoen and Howe 2008.
103 Dent 2005.
104 Perfecto 1995.
105 Shennan *et al.* 2005.
106 Shennan *et al.* 2005.
107 Hillel 1994, 224.
108 Wall and Smit 2005.
109 Wall and Smit 2005.
110 Harlan 2008.

5

Saving the soil

Soils are great! You got *inputs*, you got *outputs*, and in between you got PROCESSES!

Clare Wilson[1]

Origins of agriculture

For the first 200,000 years of our existence as modern *Homo sapiens*, human beings depended for their survival on hunting, fishing and gathering wild plants. Then, about 11,000 years ago in Mesopotamia, a small but radical change took place. In evolutionary terms, the change happened with startling rapidity. People took up farming in what has been called the 'Neolithic Revolution', a change in subsistence practice or economy that has affected almost everything about the way that we live. In order to understand just how new this economy was, it is important to understand what went before: the Palaeolithic and Mesolithic periods, when people made their living by hunting and gathering wild foods.

Hunter-gatherers usually live by migrating along with their food sources, setting up seasonal camps in different regions as they follow the fish and wild game, and seek out berries, tubers and seed-bearing plants as they become available. In a few coastal regions, prehistoric foragers were able to settle down in permanent settlements, because there were a number of different species migrating past the settlement at different times of the year, so there was no need to follow any particular animal on its migration route. Typically, though, hunter-gatherers are more nomadic; they tend to have few possessions and few children, partly because of the need for mobility and partly because the natural landscape can only support so many people living this lifestyle. The advent of farming allowed people – or forced them – to adopt a more settled way of living. Farming increased the 'carrying capacity' of the land, *i.e.* the number of people that it can support, and the population exploded.

The Neolithic Revolution began around 11,000 years ago (*c.* 9000 BC) in Mesopotamia, which is now southern Iraq. We think that the impetus for the adoption of farming was a change in the climate.[2] Towards the end of the last Ice Age there was a warming, but then came a recurrence of cold temperatures during a 1200 year 'blip' called the Younger Dryas; this period brought droughts to the Middle East around 12,700–11,500 years ago,[3] and the drought meant that the wild resources people depended upon grew scarce. Some of the most important wild foods for these hunter-gatherers were the edible seeds from a vast range of wild grasses, including the ancestor to today's wheat, called einkorn.[4] The drought diminished the extent of the alluvial floodplains on which these grasses grew, and so impelled people to begin the process of artificial irrigation. We think that this began with people broadening the floodplain in order to encourage the growth of edible grasses.[5] Gradually, the form of these grasses changed and developed as people selected particular plants and varieties for their useful characteristics, and so wild grasses evolved into domesticated wheat and barley, which are quite distinct from their wild ancestors. The domestication of wild animals happened a few hundred years later, beginning with sheep and goats.[6]

The domestication of plants and animals has affected every aspect of human behaviour, and it has had an immeasurable impact on the scope of human accomplishments. Agriculture enabled people to come together into larger, permanent settlements. It enabled the population to increase exponentially, and as people came together in the first towns, it enabled a more complex type of society to develop, with specialist crafts and a more complex social hierarchy to control both land and people. Writing was initially developed in order to keep track of produce and trade, so the first records we have are rather dry logistical notes, but they give us an insight into how people lived and how land was managed around 5000 years ago in ancient Mesopotamia.

Mesopotamia – the land between the Tigris and Euphrates Rivers – is also called the Fertile Crescent, because of the abundance of water and good quality soils. The region was cool and dry during the last glaciation, but around 9500 BC the climate became warmer and wetter.[7] People undertook a sort of pre-farming phase by artificially expanding the floodplains so that more wild grasses could grow there; the edible seed heads provided protein, and included the forebears of today's cereal crops. As the process of domestication of cereals progressed, temporary irrigation channels were excavated, and in time the temporary channels of the early period were replaced with more permanent irrigation structures.[8] As agriculture grew more intensive, massive mud brick sluices or 'regulators' were built on the Euphrates to regulate the flow of water out of the main river channel into the arable fields; one ancient text describes a regulator for a canal being built from 648,000 mud bricks – a massive endeavour.[9] Managing these systems was so important that villages or districts had officials who were in charge of irrigation matters, including rotation of the water between different fields, building and repairs, and clearing silt from the channels.[10]

By 4500 BC all the best land in Mesopotamia was being cultivated, and ploughs were introduced to increase the scale of production.[11] Towns grew into cities, and by 3000 BC the population of the city of Uruk, for instance, had reached about 50,000.[12] It has been estimated that almost two thirds of the 35,000 square miles of arable Mesopotamian land was under irrigation when the population reached a peak of about 20 million people.[13] In 3000 BC, the period of peak population, the records show that there were equal amounts of wheat and barley being grown in the fields, but during the next thousand years the soil quality declined.[14] As the population grew, the fields had been left fallow for shorter and shorter periods of time, and over time this compounded a fundamental problem: salinisation of the soil and water.

It is as important to drain water off the fields as it is to bring water into them, because water in arid regions tends to be salty, and salt is toxic to many plants. All rocks contain salt, and when rocks break down they release this salt into soils and groundwater. In temperate regions the salt is washed away in streams, so that much of it makes its way out to sea, but in arid regions it poses more of a problem. Salt builds up in the groundwater, and it is therefore important that the water table is kept well below the soil surface so that it does not damage or kill the crops. Over-irrigation raises the level of the water table, and evaporation pulls the water up through the soil, where the salt can precipitate out onto the soil surface, forming a toxic white crust (Figure 5.1).

Figure 5.1: Bonneville Salt Flats. (Photo © Berzina/Shutterstock)

We know that salinisation occurred in ancient Mesopotamia, because Babylonian texts refer to the whiteness of the fields when they describe famines, and Sumerian surveys refer to salt in the soil.[15] Records show that wheat yields declined until around 2000 BC, when the Mesopotamian farmers stopped growing it altogether.[16] Barley is a more salt-tolerant and resilient species, and by 2300 BC[17] it came to replace wheat altogether. At the time, the climate in the region was becoming hotter and drier, which caused even more stress on the crops.

The increasing aridity may have been the main catalyst for change, so the problem of salinisation may not have been due simply to over-irrigation. The period from 2400–1200 BC was a period of significant climate change, with glaciers melting at an increased pace, sea levels rising, and a consequent rise in ground water level. In the temperate zones the warming temperatures led to an improvement in agricultural capacity, but in the Mediterranean and Middle East the climate became drier. The increased aridity would have compounded the salinisation problem, especially as the groundwater rose. Water harvesting went some way towards ameliorating the problems of water shortage, but in 1300–900 BC the agricultural system finally collapsed and many cities were abandoned, their wells having gone dry.[18]

Today, irrigation is the mainstay of modern agriculture, with 71% of pumped water going for agricultural use. Salinisation is now a global problem, and more than 10% of the world's irrigated land is suffering from the effects.[19] The UN's Food and Agriculture Organisation (FAO) estimates that 50% of the world's irrigated land is currently suffering from salinisation. The extent of irrigated land is even thought to be declining, because of this degradation of the soil. Some arid areas are not affected, because of the good drainage, but in most arid regions the sub-soil is too impermeable for it to drain effectively. To compound the problem, tube wells have lowered the level of the groundwater in the aquifers, which might seem like a benefit but in fact it has had the reverse effect. To maintain their purity, aquifers need water coming in as well as water flowing out – and tube wells have lowered the water level to the point where the rain can no longer permeate down to a sufficient depth.

Sustainable irrigation

There are, however, ways to practice sustainable irrigation, with fallowing the land probably being the oldest practice.[20] Land can be fallowed with plant species that accumulate salt, and the plants can then be removed and burned or disposed of in such a way as to prevent salt from returning to the soil. Terracing the hillsides also seems to prevent salinisation, either because of the steady movement of water through the system, which might serve to wash out any salt accumulations, or because of the regular addition of fresh sediment to the terrace soil surface.[21] An archaeological investigation of an extensive 15th–18th century terraced landscape in Tanzania has supported this finding; the geoarchaeologist found microscopic evidence for wetting and drying, but they found no evidence for salinisation in these formerly irrigated fields[22].

Another approach to salinisation is to grow crops that use less water, or to grow crops that are tolerant of salinity. One new study has provided excellent results by growing salt accumulating species together with a salt tolerant plant that can be used for cattle fodder.[23] In one experiment, 60,000 tons of animal fodder was produced per hectare in Balochistan, an extremely arid region of Pakistan.[24]

Another relatively new approach to irrigating the desert is drip irrigation, which delivers water in very small quantities, directly to the roots of the crops. This system saves water, limits weed growth, reduces the likelihood of waterlogging and salinisation, and it puts less stress on the crop. It could be regarded as a modern form of clay pot irrigation, which I described in Chapter 3. Drip irrigation has recently been improved by the invention of an irrigation system made from a plastic that retains salt and industrial pollutants, but which allows water to permeate through the porous pipes.[25] There are no filters to become clogged, and cleaning the system simply involves periodically flushing the pipes with fresh water. The system saves water and functions much like Fan Sheng-chih Shu's buried pot system, with the plants drawing up water as they need it from the buried pipes, which release water into the soil as it is needed. Experiments in the UK and US have proved to be very successful, and it is now being tested in Chile, Libya, Tanzania, Mauritius, Spain and the Middle East.[26]

Anthropogenic (man-made) soils

In marginal regions with extreme climates, the weather, topography or poor soils can make it almost impossible to grow crops – but our ancestors found all kinds of ways to live in such inhospitable places. Often they didn't have a choice. People were often forced onto poor land because of population growth, *e.g.* in North America the American Indians initially cultivated the rich valley soils, but over time people moved onto areas of poorer soils as more and more farmland was needed.[27] This process occurred all over the world, wherever populations expanded; in both historic and prehistoric times there are many examples of populations being pushed onto poorer land. Early farmers responded to the difficulties of farming marginal soils with common sense and ingenuity.

In the Netherlands and Germany there are vast areas of sandy, free draining soils that are difficult to farm because they do not retain enough water or plant nutrients for crops. In the Middle Ages, farmers developed a way of artificially deepening and enriching these soils by cutting turves of sod or peat from uncultivated land and using it as bedding for cattle and sheep. When this bedding was removed, the manure-soaked turves were composted and then spread onto the fields to enrich the soil, adding much needed nutrients and organic matter.[28] Over time, this created a very deep, rich topsoil called a 'plaggen soil', from the German word 'plagge', which means 'sod'. The same system was used in Scotland in the Late Norse period of the 12th–13th centuries, after the invading Vikings had settled down to become farmers.[29] The realisation that 12th century Norse farmers were creating plaggen soils was extraordinary enough, but then in the 1980s an even more surprising discovery was made: deep, rich topsoils were discovered in and around prehistoric sites. Were these ancient soils created in the same way as the medieval ones? My PhD was based on this question – and so I ended up working in Orkney and Shetland, the beautiful, remote and windswept islands that we call the Northern Isles of Scotland.

The Orkney Islands are known for their Neolithic monuments – massive stone tombs and standing stones, and Stone Age villages such as Skara Brae and the Ness of Brodgar. The archaeological remains are so impressive that the centre of Orkney is now a World Heritage Site. These Neolithic villages are unusual, because settlements in this region at this time were more often made up of just a pair of houses, perhaps representing one dwelling with an annex where cooking or other activities may have been carried out. Neolithic houses were often built into great piles of ash and other rubbish, known to archaeologists as 'middens', but these middens are also found spread out around the settlements, sometimes covering an area measuring tens of metres, but sometimes covering more than a hectare of land.[30] Until recently, it was unclear why the middens were spread so widely, and why they were not used as fertiliser on the fields.

Another unexplained phenomenon was the occurrence of plough marks that were found underneath the midden material on several sites. We know that the Neolithic people were growing wheat and barley, because we find the charred seeds on the sites – but it appeared to archaeologists as if people were removing the topsoil to use elsewhere, and dumping rubbish onto what was formerly agricultural land. On one site it was suggested in a preliminary analysis that the soil might have been removed first, as a valuable commodity, before the rubbish was dumped.[31] It has also been suggested that the midden material may have had some ritual meaning, given that it might represent fertility.[32] I wanted to see if there was a rational explanation rather than a ritual one, and so I carried out microscopic and geochemical investigations into the soils and middens on two well-dated prehistoric sites.

The study of the microscopic structure of the soil is called soil micromorphology, and it is carried out by taking undisturbed soil samples in columns or in tins that are carefully eased into the exposed soil face section (Figure 5.2). The moisture is removed from the sample using acetone, and then it is set in resin, so it forms a solid block that can be cut into thin 'slices'. One 'slice' is then bound to a glass slide and ground down to 30 microns, which enables us to look through it using a polarising microscope (Figure 5.3). This enables us to study the structure of the soil, so we can identify the features produced by earthworm activity (this is useful, because plenty of earthworm activity tells us that a soil was reasonably fertile), and we can see evidence of disturbance, and we can also see things like bone fragments, charcoal, ash and plant remains that either accrued naturally or were added to the soil by people. We can also look at the soil chemistry, and taking samples for geochemistry adjacent to the micromorphology sample enables us to test the levels of plant nutrients such as nitrogen and phosphorus.

Figure 5.2: The author taking micromorphology samples. (Photo © Jerry Bond)

Figure 5.3: Thin section slide. (Photo © the author)

In many of my analyses I also look at the soil magnetism. All material has some magnetic charge, although it can be ever so slight, and the increased charge of certain soil samples provides distinctive clues for archaeologists. Topsoil has a stronger charge than the lower soil layers, because a charge is produced by soil bacteria, so ditches and other features that have been dug out and silted back up again are identifiable using survey techniques that measure soil magnetism. If you've ever watched 'Time Team',[33] you'll have seen them carrying out the 'geophys' – archaeological shorthand for the various geophysical surveying techniques that include magnetometry. Another material that has a strong magnetic charge is ash, which is useful in archaeological studies because it is plentiful on settlement sites. Ash accumulates in houses and yards where people live and cook, but it was also an important fertiliser in the temperate northern European countries, because it neutralises our often rather acidic soils, rendering them much more fertile. I compost it myself, and add it to my vegetable and flower beds.

For my research, I analysed midden deposits in Orkney and Shetland, and compared the middens to the soils that lay beneath them. I began by investigating a farm that was known to originate in the Late Norse period (12th–13th century), based on both radiocarbon dates and the fact that it had the Norse place name of Bragasetter, which comes from the old Norse 'brekka', meaning 'slope', and 'setr', meaning a dwelling place. There is also a legal record regarding this farm, dating from AD 1299.[34] The farm is on a small island called Papa Stour (Island of the Priests), which lies off the west coast of Shetland.

From the Middle Ages right up until the 1960s, the farms on Papa Stour had been managed using the traditional plaggen system, which was identical to that of Germany and the Netherlands. Turves of peat were cut from the wild and rugged moorland on the western part of the island (the 'outfield'), and used as bedding for the cattle and sheep kept in the byres. Every spring, the bedding would be spread onto the infields, which were divided from the outfields by a large turf wall. Over time (about 800 years, to be precise), the deposition of these manure-soaked turves of peat created a deep, rich soil up to 1.20 m deep. The rich green of the grass on the much-improved infields on Papa Stour were quite distinct from the yellowing vegetation on the poor, thin, stony moorland of the outfields, from which most of the peat had been stripped (Figure 5.4). This system lasted at Papa Stour right up until 1967, because the remoteness made importing chemical fertilisers problematic, but also because the existing system was so effective. When I was doing my research, the farmer who had worked the land was still living at the farm, and was able to tell my team about how the system had worked when he was young. He explained that each family had an area of land in the outfield from which they cut their turf, and each also had rights to a stretch of beach from which to collect seaweed – another excellent fertiliser, as it is high in nitrogen.

When I carried out my study, I took soil samples from all the different parts of the farm, including the deep, rich soils of the 'kaleyard', the walled garden where

Figure 5.4: Bragasetter farm. (Photo © the author)

Orkney and Shetland farmers grow vegetables that need shelter from the powerful and incessant winds (Figure 5.5). I took samples from the fields near to the house, and from the more distant fields, and from the 'planticrues', the little stone enclosures where seedlings are grown; I also sampled the infield grazing lands, and the rough outfield moorland. The latter was for 'control'; I wanted to have a baseline of unaltered soil against which I could measure the characteristics of the amended farm soils.

My results demonstrated clearly what archaeologists call the 'shlepp effect' – the notion that people will only carry things as far as is necessary. Thus, the inner fields and garden had the richest soils, with less and less material being hauled out to the more distant fields surrounding the farmhouse, byre and barn.[35]

I was also able to identify all the microscopic characteristics that make up a plaggen soil. The samples I looked at contained many microscopic fragments of peat, just a few millimetres across at most. Some of the peat fragments had been charred, and some were not, reflecting the fact that some derived from hearth sweepings and some from the bedding in the cattle byres. I could also see bone fragments, although they were very decayed. The fertilisers I could see under the microscope were the very ones that the farmer described having added – although sadly, there was no way to identify the added seaweed. It decays thoroughly, and leaves no marker apart from the shells of sea creatures that cling to it. These shells can be quite distinctive and are a good indicator for added seaweed, but this will have to be another project.

Figure 5.5: Sampling at Bragasetter Farm, Papa Stour, Shetland. I am taking soil samples, while my colleague Paul is recording the exact colour using a Munsell Chart. (Photo © Richard Evershed)

Having worked out how to identify a medieval plaggen system, my next step was to investigate two multi-phase settlements that had been buried by sand on several occasions. I began with Old Scatness, an important Iron Age village in Shetland, which was being excavated by Steve Dockrill from Bradford University. I was intrigued by the plough marks filled with red midden material, and by the ash midden that lay over the plough marks (Figure 5.6). I took samples from the whole sequence of soils, from the modern day and through the Norse levels and three phases of Iron Age soils, and down to the Bronze Age. When I looked under the microscope, I made an extraordinary discovery. You can see the excavation of the plough marks filled with this very red material; peat ash can be this very distinctive red colour in the field, although sometimes ash can only be recognised under the microscope. The remarkable thing was that when I looked under the microscope, the agricultural soil was indistinguishable from the ash midden. Compare the photos: Figure 5.7 shows a thin section sample of the red layer that you can see being excavated in Figure 5.6, magnified 25 times. Figure 5.8 shows an early Iron Age midden from another part of the site, also magnified 25 times. In both samples you can the bright orange of the peat ash, and you can also see dark shapes with mineral grains embedded in them. Those are fragments of charred peat. These samples show that the Bronze Age to Iron Age soil and the Iron Age midden are made up of *exactly the same material.*

Figure 5.6: Ard marks, Old Scatness, Shetland. (Photo © Val Turner)

Figure 5.7: Bronze Age to Iron Age soil at Old Scatness. This has been magnified 25× and is under 'oblique incident light', which means I'm bouncing light off the surface of the slide. (Photo © the author)

Figure 5.8: Iron Age midden, Scatness – also magnified 25× and under oblique incident light. (Photo © the author)

I then went on to investigate a Neolithic to Early Iron Age site at Tofts Ness, on the beautiful, low-lying Orkney island of Sanday, where great clouds of seabirds flock to the deserted sandy beaches (if I'm sounding a bit lyrical, it's because you really have to visit this place to believe it – it is of surpassing loveliness) (and surpassing windiness). For my project, in 1998, I re-excavated the trenches excavated some years earlier by Steve Dockrill, because Steve had already established the dating and sequence of soils that I wanted to look at.[36] The settlement site itself was a low mound made up of middens and structures, and the structures were surrounded by an area of deep, man-made soils. I wanted to know whether they were plaggen soils, or whether they were created by some other means. Either way, they were not 'natural' to the area.

I was delighted, first of all, to discover Neolithic plough marks beneath a Neolithic soil. The next exciting discovery came later, in the lab, when once again I found that – as at Old Scatness – the Neolithic soil was exactly the same material as the Neolithic midden that formed the layer above. I then looked at the phosphates – and the results were thrilling (Figure 5.9). You don't have to be a specialist to see that this plot shows a comparatively low level of phosphate in the topsoil, with slightly higher levels in the Bronze Age. It is the Neolithic phase that confirmed my findings from the microscopic study: the Neolithic midden and the underlying soil have very high levels of phosphate and almost exactly the same organic/inorganic signature. The phosphates confirmed the evidence from the micromorphological analysis, and the evidence of ard marks that I found in the field. What this was telling me was that people on this Neolithic site had built up a midden, and then run a plough across it.[37]

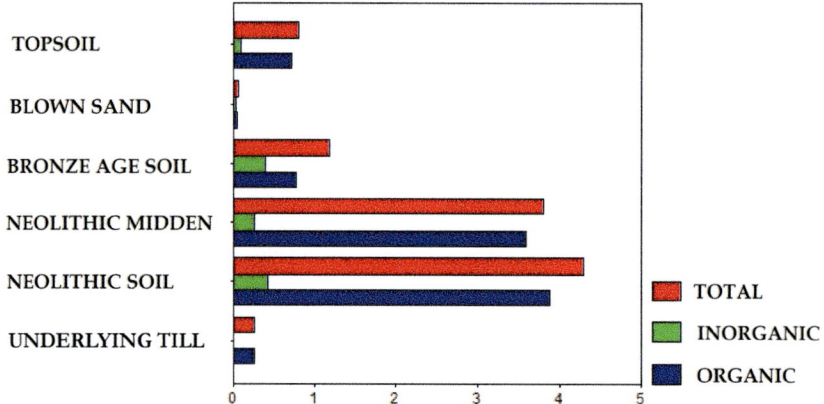

Figure 5.9: Phosphates from Tofts Ness, Sanday, Orkney. Each sample size was half a milligram of soil. (From Guttmann et al. 2004.)

My investigation set out to characterise the soils of the medieval to modern plaggen system, and to compare it with the prehistoric soils that archaeologists were finding in the Isles. I discovered that farmers in the Neolithic and Bronze Age were not spreading their compost onto their fields, because they were actually cultivating the compost middens themselves. They were creating large piles of compost and then ploughing over them, spreading them out and creating very fertile gardens on what would otherwise be sandy and unstable soil.[38] These midden soils – or perhaps we can call them garden soils – are found on Neolithic sites all over the Orkney islands, and the ancient soils are still deep, rich and fertile today.

The prehistoric practice of stabilising sandy soils with midden material also occurred in the Western Isles, where the sandy soils are called 'machair'. Here, the practice dates back at least as early as the Bronze Age, though there is also evidence that Neolithic people ploughed and cultivated the middens left by the earlier, Mesolithic hunter-gatherers.[39] Geographers used to find these humic topsoil surfaces buried in the sand dunes, and they assumed that these soils represented periods of climatic stability, when the storms and high winds that are a feature of this region settled down and allowed soil to form on the sand dunes. Microscopic investigation of the soils showed that actually, this is not always the case. Under the microscope, some of these soils have been shown to be man-made, created by dumping composted material onto the unstable sand to create small pockets of farmland on these windy islands.[40] Our ancestors turned sand dunes into gardens.

We now know that adding manure and seaweed to these sandy, windblown soils is in fact essential for their stability and maintenance. Both materials add organic matter and so create aggregates in the soil; organic matter supports earthworms, microbes and fungi, all of which also improve soil stability. Seaweed adds nitrogen and potassium, and it also contains polysaccharides which increase the soil 'stickiness', while animal manure adds all three of the key plant nutrients, nitrogen, phosphorus

Figure 5.10: Sandy field with seaweed, South Uist. (Photo © the author)

and potassium. Chemical fertiliser on its own just washes straight through the sand, because there's not enough organic material to bind it in place. The sandy environment makes organic additives particularly important, because the vegetation on the machair sands forms a thin mat which is actually very fragile. Once it is damaged, *e.g.* by the plough or by the hoofs of grazing animals, the wind can create 'blowouts' which strip away the sand and leave great craters in the fields and sand dunes. Because of this, Scottish law stipulates that farmers on the machair must apply at least 40 tonnes of seaweed per hectare, when available, or 25 tonnes of animal manure (Scottish Statutory Instrument 1993, no. 3149), and ideally they use both (Figure 5.10). In short: farmers in the Western Isles are required by law to follow the prehistoric methods of soil preservation.

Terra Preta soils

Terra Pretas are man-made soils that originate in prehistoric South America, and these extraordinary soils actually have the potential to combat global warming. Terra Preta soils are deep, nutrient-rich, highly organic and very black soils that occur as pockets within a landscape of otherwise nutrient-poor soils. One key characteristic of these soils is an abundance of prehistoric pot sherds, which indicates that they were intensively dug over by prehistoric peoples. These soils have long been well-known to local farmers in Brazil, but they caused a stir when they were discovered by geographers in the mid-20th century. Initially they were thought to have natural origins, *e.g.* as volcanic ash deposits, or perhaps as former lake beds. Such pockets of naturally fertile land (so the theory went) would have attracted early farmers, who might have added the profusion of pot sherds together with organic midden material when they began to cultivate them.[41] This idea had to be discarded, because further research indicated that Terra Preta soils were in fact man-made – with the earliest being around 7000 years old.[42]

Terra Pretas are found throughout the Amazon basin, but are mainly located on low hills along river valleys, which are also the chosen location for settlement by Amazonia's prehistoric population.[43] Studies of Terra Pretas – also called 'Amazonian Dark Earths', or ADEs for short – have shown that the soils are associated with prehistoric settlements, but most aren't simply formed by the accrual of settlement rubbish that accumulates around villages. A small proportion of ADEs are located around the actual prehistoric settlements, but most are in areas of managed land, forming fields of up to 120 ha. A remarkable characteristic of these soils is that they actually spread downwards, possibly because of earthworm activity. These soils contain up to three times more organic material than the surrounding soils, and they are also high in nitrogen, phosphorus, potassium, calcium and magnesium, which must derive from the addition of various fertilisers such as kitchen waste, excrement, plant remains and perhaps aquatic plants.[44] Significantly, the ADEs also have up to 70 times more charcoal than the surrounding soils.

The charcoal is perhaps the most important characteristic of the soil, and may hold the key to what makes it so fertile. Simply adding organic material doesn't seem to go very far towards improving fertility, because the organics oxidise very quickly and vanish into the atmosphere as CO_2 – but charcoal doesn't oxidise, and adding charcoal increases the productivity exponentially. Studies of the soils show that the charred material ranges from completely charred to only partially charred. The incompletely burned charcoal is called 'bio-char', and it has a porous structure that holds water and also binds organic nutrients. This bio-char seems to increase the soil microbe population, and it also changes the type of microbes in the soil, which may be another key to the high fertility of the soils. Field experiments in the Amazon basin showed that adding charcoal to the soil can increase crop production by up to 220%, with one experiment showing an 800% increase in rice yields.[45] Another experiment showed an increase in uptake of nitrogen by plants of up to 400% in charcoal-enriched soil.[46] Bio-char also seems to boost the effects of conventional chemical fertilisers, increasing yields by 25–50% in one study.[47] It is also a remarkably stable material; it does not break down or decay, but rather it remains in the soil, intact, for hundreds and possibly thousands of years.

These findings are relevant not just to archaeologists, but to agronomists as well. Around 3.4% of the world's population live in the tropics,[48] and improving the soils of these regions would be hugely beneficial to farmers. Agronomists are carrying out experiments that involve charring different types of locally available materials, in order to help farmers to develop local solutions to increase fertility and yields.

Bio-char has yet another benefit to the environment: it has been estimated that it could remove 1.2 billion metric tons of atmospheric carbon per year, and so it could be used to combat global climate change. Creating bio-char also produces heat, and so the process could also be used to create energy. A meta-analysis was undertaken that pooled together many different bio-char experiments,[49] and it found that 50% of the experiments studied reported an increase in crop yield following applications of bio-char.[50] There have been a number of studies that indicate that bio-char also reduces methane emissions and nitrous oxide emissions from the soil.[51] It may also help to clean up contaminated land, as both organic and inorganic pollutants – including pesticides – bind themselves to bio-char.

Soil creation and preservation in Africa

Many of us Baby Boomers remember the famine in sub-Saharan Africa in 1973, in which over 100,000 people perished. My generation was more or less force-fed more food than we wanted, with the injunction that we should be grateful to be uncomfortably overstuffed because 'children are starving in Africa'. The cultural lesson that was shoved down our throats was that Africans were eternally starving and in need, and it was our responsibility to save them – but another cultural lesson might have been that Africans have admirable traditional knowledge and resources that can teach us all some lessons in conservation and eco-friendliness. The history of the famine is interesting, as it demonstrates the problems that ensue when large-scale changes in land use are implemented without first understanding the landscape ecology.

The traditional farming and grazing practices that took place in sub-Saharan Africa were well adapted to the environment, which is not to say that there was a perfect idyll in which no famine ever took place, but simply that a working system was disrupted by European colonial governments who had different concerns, and who were also oblivious to the ecological benefits of what had taken place before their arrival. Here is the tradition of one particular region, the West African Sahel: pastoralists used to graze their cattle on the farmers' crop stubble, and the cattle manure added nutrients to the farmers' fields.[52] The herders then drove their stock north, following the fresh grass that followed the rains. When the limit of fresh grass was reached, the herders returned to the south, again grazing on farm stubble and thereby applying the precious herbivore manure that not only fertilised the fields, but which also provided vital organic matter that binds the soil and stops it from blowing away. A key element to this system was fallowing for decades at a time, and not overgrazing the land.

The French restructured the traditional land use system, which caused considerable soil degradation and erosion. The French reorganised the land on European principles, erecting fences that stopped the annual cycles of movement that had previously kept the land fertile.[53] The pastoralists could no longer move freely, and the farmers did not get the annual input of cattle manure. To compound the problems, fallow periods were reduced and the population rose, and the consequences for the soil – and thus for the people – were catastrophic.

There are in fact hundreds of traditional local solutions to soil and water conservation in Africa, mostly using ridges, furrows, pits or otherwise protected pockets of land. In Burkina Faso such pits are called zaï, and this traditional method of planting is enjoying a resurgence – and is now being taken up in Mali and Niger as well.[54] The zaï are pits of just 20–30 cm in diameter and 10–15 cm deep, depending on the soil type, but typically they are created on permeable soils that are subject to drought. Between 12,000–25,000 of these pits are dug per hectare, and the soil from each zaï is placed down slope, to trap and conserve water. The pits trap dust and plant litter, both of which are useful additions to the soil. Finer sediments and organic matter both help the soil to hold water, but farmers also add manure, further improving the soil's fertility and water retention properties. The land between the pits, being quite compact and dry, does not need to be weeded, but as further pits are dug over the years, eventually this compact soil is also excavated and improved. This means that eventually the whole field recovers its fertility.

Similar methods were traditionally used in Darfur, where U-shaped ridges serve to trap surface water in much the same way as the zaï.[55] In Zimbabwe, traditional methods include leaving tree stumps in place, spreading ash onto the soil and practicing minimum soil disturbance using hoes; they also intercrop, so that the soil is never fully exposed.[56] Mulches of weeds or burned crop residues are applied to further increase soil fertility. In Ghana, grass strips between arable plots reduce erosion, and the grass is harvested for thatching, brooms, baskets and mats.[57] It is astonishing, just how much simple, non-industrial methods can do to regenerate the soil even in such arid regions.

There are many success stories coming out of Africa today – many new/old methods of ecologically sound agriculture and sustainable practice that enable smallholders to thrive. Unfortunately, there are also terrible depredations due to war and the mass movement of people as a consequence of war. It is heartbreaking to think of the wasted potential of the land and of the people in these areas. However, it is uplifting to reflect on the extensive regions where farmers are rehabilitating damaged land. It goes to show that even denuded and barren landscapes can be brought back to life.

Combating soil erosion

It has been estimated that *c.* 10% of the global land surface has been degraded by human activity, with forests being cut down and transformed into grassland, and grassland – unfortunately – being transformed into desert.[58] Soil erosion is causing the loss of a further 6 or 7 million hectares of land per year, and in 2007 it was estimated that 24 billion tons of soil are lost to erosion every year, a process that has been succinctly described as 'skinning our planet'.[59] In 2012, this figure was estimated at 35.9 billion tons per year, so the problem is getting worse, not better.[60] This has been described as 'soil abuse', and this major environmental concern has been evoking a considerable amount of attention from governments, environmentalists, agronomists, NGOs and the UN. The UN recently estimated that 33% of the world's agricultural land has been degraded.[61] Clearly, we need to act!

Most methods of farming disturb the soil, breaking up the natural aggregates (called 'peds') and exposing it to erosion by wind and water. Grazing animals break up the aggregates with their hooves, so overgrazing pasture land causes erosion when the grass or other protective vegetation is broken up. As soils become shallower, there is less depth for plant roots, which makes it harder for plants to absorb moisture; soil on sloping ground is therefore subject to droughtiness (*i.e.* it can't hold enough water). Exposing the soil surface by ploughing causes similar damage. The rain actually breaks up the natural peds, which causes a number of problems. First of all, it causes the sand, silt, clay and humic compounds that make up these aggregates to wash down into the soil structure, clogging up the natural pores in the soil and starving it of oxygen. Secondly, it causes soil to wash down hill slopes – and once it is set in motion, it becomes sediment. Sediment eroded down hills is called 'colluvium', and we find it accumulated in the bottom of dry valleys. However, when there are rivers or streams at the bottom of slopes, the colluvium washes into the water, and it then becomes alluvium, which is waterborne sediment. Alluvium can choke up the river, and when the force of the water slows, it deposits the alluvium on the river bed, the river banks or on flood plains. The accumulation of sediments in river beds raises the level of the water, making the floodplains more vulnerable to overbank flooding. All this disruption to the ecosystem takes place because of the plough.

Soil erosion in the UK can be traced back to the start of the Neolithic, around 6000 years ago.[62] In the European Neolithic, the first farmers developed a primitive kind of plough called an 'ard'. Ards did not turn the earth over, but they disturbed it, breaking up the surface vegetation and creating a furrow in which to plant crops. The resulting erosion caused the build-up of alluvium and colluvium, which in Britain we can radiocarbon date to about 4000 BC.[63] Ploughs that actually turn over the soil, called 'mouldboard' ploughs, were developed in the Middle Ages (or possibly earlier, in the Anglo-Saxon period), and the advantage was that the improved plough destroyed weeds and created a good seed-bed for planting. The disadvantage is that it increased erosion exponentially, especially when ploughing was directed up and down slopes – but erosion can be combated by ploughing with the contours, and by creating field boundaries.

We start finding field boundaries in the Neolithic on Continental Europe, Ireland and Scotland, and in the Middle Bronze Age in England.[64] Field boundaries provide a check to soil erosion, with eroded soil building up against banks or hedgerows which prevent or at least limit the process of soil erosion into valleys and rivers. In Britain, rivers began to silt up with alluvium due to ploughing in the Neolithic,[65] but by the late Bronze Age we see extensive field systems covering large areas of land in our river valleys – now buried under meters of alluvium.[66] It is worth noting that up to 95 tonnes of soil per hectare can be lost to erosion when hedges and field boundaries are removed,[67] so whether the early farming population knew it or not, these boundaries served an important function.

In the Iron Age in Britain these field systems seem to have been abandoned, and instead we have distinct landscape features called 'lynchets'. These are banks of soil that form ridges along the contours of hillsides, either marking the lower edge of ploughing, or perhaps having formed up against long-vanished walls or hedges.[68] Either way, the soil was preserved to some degree. This is significant, because Iron Age agriculture in Britain was intensive. At this time the population grew substantially,[69] and we also have written records from the Roman Empire recording that the Iron Age people of Britain were exporting grain to the tribes of Gaul. The technology that supported agriculture also improved, including iron tips on the ard ploughs, iron scythes, and rotary querns to speed up the process of grinding grain into flour.[70] The use of fertiliser also increased and intensified, with more animal manures being used at this time.[71] Small prehistoric fields are today called 'Celtic fields', and although this is a misnomer (most are actually Bronze Age), even so it is significant that the small fields that typify early British agriculture are more effective at preserving soil from erosion than large modern ones, especially when they are on slopes.

Europe is unusual in having this history of ploughing; in most of the rest of the world, primitive agriculture involved the use of digging sticks, which were used to poke a hole in the ground. The farmer would then drop a seed or seeds into the prepared hole, cover it with loose soil, and compact the seedbed by stepping on it. We now have a modern, mechanised approach to agriculture without the plough: no-till or 'conservation tillage' agriculture.

Fighting soil erosion: no-till or conservation tillage

The ard plough first appears in Europe around 5000 years ago. This primitive plough cut through the soil without turning it over, and it was not until the Middle Ages that the mouldboard plough – which actually turned over the soil – became widespread. It was in the Middle Ages that Britain also developed a form of land use called 'ridge and furrow', in which farmers owned strips of land within large, communal fields. Planting took place on the ridges, where presumably the added depth of soil was beneficial to the crops. The fields had grass strips between them, which (so I have heard) is apparently the best habitat for harbouring the insects that predate crop pests (for some reason, grass strips are better than hedgerows).

Ploughing breaks up weeds and creates a fertile seedbed for planting. The soil is rolled or harrowed to break up the clods of earth into a fine, crumbly tilth. Unfortunately, ploughing also exposes the soil to erosion, partly because the soil is exposed to wind and water erosion, and partly because the fine tilth isn't the same as the natural soil aggregates, called 'peds', that are the building blocks of a healthy soil. The benefits of ploughing were taken for granted until the publication in 1943 of Edward Faulkner's great work, *Plowman's Folly*. In it he argued that ploughing is in fact terrible for the soil, and he experimented in 'no-tillage' field trials. His first experiment proved to be a huge success, and thus the concept of no-tillage or conservation tillage was born.

The idea of no-tillage is that the soil should be disturbed as little as possible. The crop residue from the previous year is neither burned off nor ploughed back in, but rather it is left as a mulch on the soil surface; at least 30% of the soil surface should ideally be covered by this organic residue.[72] Seeds can then be planted directly into the decayed remains of the previous year's crop (zero tillage), or a slot can be cut into the otherwise undisturbed soil, and seeds can be planted directly into this slot.[73] This protects the soil from erosion, which can be reduced to near zero,[74] and it conserves moisture and increases the soil's organic matter content – which increases the earthworm and soil microbe population. Increased soil organic matter also increases the soil's nutrient status, and it helps it to hold water. It also sequesters carbon that would otherwise be released into the atmosphere. On top of all these benefits to the environment, it also reduces the number of tractor journeys over the soil, which has another set of benefits, including a lessening of soil compaction and reduced fuel usage.

Conventional farming involves 5–10 tractor trips over each field, including ploughing, harrowing, rolling, sowing, fertilising, spraying the weeds, spraying for pests and occasional subsoiling (deep ploughing to break up hardened layers of soil just below the ploughsoil).[75] With no-tillage, just one to three tractor passes are needed – one for planting, one for spraying and one for subsoiling (if the ground becomes too compact). The reduction in tractor use means the farmer uses up to 80% less fuel, and cuts working hours by 60% – and after two years, the soil becomes much less compact.[76] The soil has better structure and there is less water runoff and substantially less erosion – often near to zero. The increase in soil organic matter means that the soil has better water retention, which means lower irrigation costs, and water conservation benefits the local environment. Less water and soil runoff mean less water pollution, and the increased soil organic matter also has a chemical reaction in the soil that makes plant nutrients more available to the crops. This reaction is called the cation exchange capacity of the soil (CEC), and a high CEC means that the nutrients (which are positively charged) can be more readily absorbed by plants. Increased soil organics can increase the CEC by up to 70%.

Lower tractor use means the farmers use less fuel and produce much lower emissions, so switching to no-tillage reduces net carbon emissions by 31 kg per hectare per year, which is great for the environment and also saves farmers around 50% of their production costs, although this varies depending on the crop.[77] In short: everybody wins!

By 2017, conservation tillage was being used on 180 million hectares,[78] mainly in the Americas, with Brazil, Argentina and Paraguay producing half their food using no-tillage.[79] Over 16 million hectares in Europe have gone over to no-tillage, including

17% of the arable land in France and 20% in Germany. It has been estimated that if Europe went over completely to no-tillage, it would offset ALL the fossil fuel related carbon emissions produced by our agricultural activity.[80] The Incas, Egyptians and many other early farmers used no-tillage systems,[81] and we are now able to do the same – but on an industrial scale.

Agricultural terraces

Although we are now losing 75 billion tons of agricultural soil to erosion every year, we have the technology to conserve most and possibly all of this precious resource. Building agricultural terraces is an ancient conservation method that can preserve 90% of the soil on hillslopes.[82] Terraces also allow the soil to retain 95% or more of the rainwater that would otherwise percolate downslope or race downhill in destructive streams and rivulets.[83] Rainfall on bare slopes causes sheetwash (erosion of the topsoil surface) or rill erosion, with small rills cutting down to form gullies. Agricultural terraces are walls built along hillside or mountain contours, and this solution is so obvious that the notion has been developed independently all over the world. Terraces can be built on dry land, for trees, grape vines and all kinds of crops, or they can be built to retain paddy fields, creating shallow lakes of water that seem to cling to the hillsides in China and the Philippines. These hillside rice paddies are found all over Asia, including Japan, Korea, Ceylon, the Himalayas and India (Figure 5.11). They are also widespread in Africa.[84]

In South America, the first terraces are thought to have been constructed by the Huarpa people,[85] who lived in the Andes in Peru from around 400 BC. The Huarpa lived in villages near rivers and springs, and they not only built terraces for agriculture, but also developed the first irrigation systems in the region.[86] Maize was a staple crop in the Ayacucho Valley by 800 BC, and by the Huarpa period people's dependence on this crop rose significantly. We know this because we find grinding stones and the remains of maize on archaeological sites, but also by analysis of the human bone from this period, which has a distinctive isotopic signature that tells us what the people ate.[87]

Terracing and irrigation developed further in the Wari period, following the Huarpa. Maize consumption rose to very high levels at this stage, and society reached the levels of complexity and sophistication that we define as State level civilisation. The Wari irrigation systems were so well-planned that the people survived a 30 year drought in the 6th century AD.[88]

The Inca developed the terraces and irrigation systems still further, and excavations have shown the careful thought that went into developing the agricultural terraces of Tipon, near the Inca capital of Cusco. The terraces were 4 m high, and behind them was a basal layer of large rocks and boulders, placed there to ensure good drainage.[89] Above these was a layer of small stones, above which was a layer of sand and gravel. The topsoil above this was deep, rich and fertile, never waterlogged but never dry, thanks to both drainage and to the elaborate irrigation systems which included fountains

Figure 5.11: Terraced rice paddies near Sapa, north Vietnam. (Photo © Blue Planet Studio/Shutterstock)

for added beauty. The Inca also built reservoirs on high ground, thus ensuring there was enough water during droughts. Fortresses were also built on high ground, and nearby reservoirs ensured there was a sufficient water supply even during sieges.[90]

The long-term benefits of pre-Columbian terracing has been demonstrated by analysis of the soil chemistry and microbial activity in the ancient terraces. One study compared the soil in three distinct areas:

a) terraces that are still cultivated today,
b) Inca terraces that have been abandoned, and
c) uncultivated local soils under native vegetation (grasses, shrubs and cacti).[91]

The sites are in the Colca Valley in Peru, high in the Andes where the climate is semi-arid. Radiocarbon dates indicated these terraces are 1300–1700 years old, and some are still in use by farmers using the traditional methods. It seems extraordinary, but the soils that have been cultivated for around 1500 years were much richer than the local uncultivated soils, with higher levels of carbon and nitrogen. Phosphorus levels were also high in the soils of the abandoned terraces, with high phosphorus levels also being maintained on the cultivated terraces, but with lower levels on the uncultivated ground. Analysis of the soil enzymes demonstrated that the organic material added over 400 years ago was *still* affecting the soil biology. It is amazing to think that 1500 years of cultivation should leave soils richer than soils that have been left alone, but that is the unambiguous result of this research. Crop rotations with legumes, plus the regular addition of animal manures, have improved these soils and demonstrated that ancient farming technologies really are effective, not just in conserving the soil but in actually improving it.

Terraces are also common around the Mediterranean, for instance in Italy, where they date back to the Neolithic.[92] The Greeks and Phoenicians built terraces around 2500 years ago,[93] and olive trees up to 700 years old grow on top of the terrace walls.[94] One site in Crete has an olive tree 4.5 m in diameter which was dated through dendrochronology to the Hellenistic or Roman period, but while such ancient trees are very rare, Greek and Roman terraces were fairly common – though rather difficult to date. On the island of Delos, archaeologists have pieced together a convincing date of origin using several methods: first of all, there is extensive ancient terracing with Hellenistic and Classical period pottery in the soil, and secondly, the construction style for the terraces is the same as that of the Hellenistic to Classical settlements with which they are associated.[95] We don't get much assistance from the ancient texts on the topic of terracing, so we depend largely on evidence from pottery and radiocarbon dating.

In 1909 an American professor of soil science and agriculture, Franklin Hiram King, took an investigative tour of Japan, China and Korea, in order to study the agricultural methods in the region. He reported that in Japan, terraces have stone walls reaching up to 12 feet high, with each terrace divided into 20–30 foot gardens.[96] There were drains or furrows along three sides of each garden to collect surface water for storage in basins within each field, and traditionally the water from the paddy fields was not released from each terrace until after the plants had had time to absorb the nutrients, so the nutrients, soil and water were all conserved. Manure was carried to these fields in baskets on bamboo poles, and boatloads of nightsoil (human excrement) was also added. The average paddy field was described as 31 × 40 feet, but 74% of the fields were smaller than 1/4 acre. In China, terraces as small as a few metres have been reported, while in Japan one field was described by a colleague of King as being 'barely larger than a good napkin'.[97] This shows an appreciation for every centimetre of land and its potential, and also a careful conservation of nutrients.

Soil: our lives depend on it

It pains me when people speak disparagingly about 'dirt'. It is equated with uncleanliness, with worthlessness and with poverty – and yet, the soil is the precious resource in which we grow our food. It also underpins the beautiful landscapes that uplift our spirits. The chemical processes that take place within it are – to me – beautiful in their steady predictability. Not everyone will feel the same devotion to the minutiae of the life of the soil, but nevertheless, all our lives depend on it. Furthermore, the soils and sediments that we archaeologists devote ourselves to excavating are filled with information about the past, and that information is gold dust. Soil science and archaeology together can tell us how our ancestors managed and conserved their land, and this knowledge can help us to build a better and more sustainable future.

Notes

1 My editor feels that this comment ought to be referenced, so I will add that Clare was joking, although there is also an element of truth to her throwaway comment. I think she said this during one of our conversations in the microscope lab when we were PhD students at Stirling, but we may have been down the pub.
2 McCorriston and Hole 1991; Hillman 2000; Gupta 2004.
3 Robinson *et al.* 2011.
4 Hillman 2000.
5 Hillman 2000.
6 Legge and Rowley-Conwy 2000.
7 Barker 2000; Robinson *et al.* 2011.
8 Postgate 1992, 178.
9 Postgate 1992, 178.
10 Postgate 1992, 178–179.
11 Montgomery 2007, 37.
12 Montgomery 2007, 37.
13 Montgomery 2007, 39.
14 Montgomery 2007, 40.
15 Montgomery 2007, 40; Postgate 1992, 181.
16 Montgomery 2007, 40.
17 Issar and Zohar 2007, 122.
18 Montgomery 2007, 40.
19 Postel 1996, 22.
20 Fallowing meaning leaving the land uncultivated for a period, to allow it to recover fertility.
21 Lang and Stump 2017.
22 Lang and Stump 2017.
23 Khan *et al.* 2009.
24 Khan *et al.* 2009.
25 Scott 2009.
26 Scott 2009.
27 Doolittle 2000, 155, 163, 164.
28 Pape 1970; van de Westering 1988; Spek 1992.
29 Simpson 1993.
30 Guttmann 2005.
31 Hunter *et al.*, unpublished. (He changed his mind about this interpretation following my research, so this preliminary theory was discarded and does not appear in the final publication on the site of Pool.)
32 Parker Pearson *et al.* 1996.
33 A popular UK-based television series that presented archaeology to non-archaeologists. This engaging program ran for 20 years, and inspired many young people to study archaeology for their degrees.
34 Crawford and Ballin-Smith 1999.
35 Guttmann *et al.* 2006.
36 The site of Tofts Ness is now fully published: see Dockrill *et al.* 2007.
37 Guttmann *et al.* 2006; Guttmann 2005.
38 Guttmann 2005.
39 Gregory *et al.* 2005; Guttmann 2005.
40 Gilbertson *et al.* 1999.
41 Woods and McCann 1999.

42 Glaser 2007.
43 Mann 2002.
44 Glaser 2007.
45 Steiner *et al.* 2004.
46 Steiner *et al.* 2003, cited in Tenenbaum 2009.
47 Tenenbaum 2009.
48 Glaser 2007.
49 A meta-analysis is an analysis of previous analyses, in which usable data from lots of different studies is pulled together into one big database and analysed all together. It is a way of increasing the sample size and thus producing more convincing results.
50 Spokas *et al.* 2012.
51 Briones 2012.
52 Montgomery 2007, 166.
53 Montgomery 2007, 166.
54 Ouedraogo and Kaboré 1996.
55 Mohamed 1996.
56 Hagmann and Murwira 1996.
57 Millar *et al.* 1996.
58 UNEP 1990.
59 Montgomery 2007, 3.
60 FAO 2017b.
61 FAO 2017b.
62 French *et al.* 1992; Scaife and Burrin 1992; see also Brown 1997.
63 French *et al.* 1992; Brown 1997; Bell and Boardman 1992.
64 Bradley 1978; Barclay 1997; Cooney 2000.
65 French *et al.* 1992; Scaife and Burrin 1992; see also Brown 1997.
66 Yates 1999; 2001.
67 Pretty 1998a, 69.
68 Cunliffe 1991, 377.
69 Cunliffe 1991, 541.
70 Wells 1984, 160.
71 Guttmann *et al.* 2005.
72 Abdalla *et al.* 2013.
73 Gliessman 2007, 111.
74 Reicosky and Saxton 2007.
75 Baker and Saxton 2007, 6.
76 Baker and Saxton 2007, 6.
77 Baker and Saxton 2007, 6.
78 Jules Pretty, pers.comm.
79 Abdalla *et al.* 2013.
80 Abdalla *et al.* 2013.
81 Derpsch 2004.
82 Montgomery 2007, 175.
83 Hillel 1991, 97.
84 *e.g.* Lang and Stump 2017. See also the World Heritage site of the Konzo terraces in Ethiopia: https://whc.unesco.org/en/list/1333 Accessed 9 November 2018.
85 Wright *et al.* 2006, 27.
86 Wright *et al.* 2006, 27.
87 Finucane 2009.

88 Wright *et al.* 2006, 27.
89 Wright *et al.* 2006, 15.
90 Mazoyer and Roudart 2006, 196.
91 Dick *et al.* 1994.
92 Tarolli *et al.* 2014.
93 Dimakopoulos 2016; Zurayk 1994.
94 Price and Nixon 2005.
95 Price and Nixon 2005.
96 King 2004, 45.
97 King 2004, 277.

6

Vernacular architecture and sustainable cities

Vernacular architecture is the architecture of ordinary, non-monumental buildings, which can include houses, farms and industrial buildings. The term was coined in 1839 and originally referred to buildings of minor importance, but it came to mean buildings constructed in local styles, often being built by people who are not professionals and often without the use of drawings or plans. In the late 19th century there was a rekindling of interest in local styles of vernacular architecture, an interest that was drawn from the Romantic Movement and the Arts and Crafts movement; essentially it was a rejection of mass-produced housing and the increasingly industrialised landscape. The movement was characterised by a certain amount of idealisation of the past, but it also focused on the timeless notion that we should live in an environment that is not only productive, but also beautiful and inspiring.

The study of vernacular architecture has been developing with renewed interest since the 1960s, with increasing recognition that vernacular architecture does not just belong to the past – it has great potential for informing the buildings of the future. Crucially to our topic of sustainability, it also refers to buildings that use local resources, and there are now an increasing number of architects as well as individuals who are interested in designing buildings using local materials, often with new energy technologies built in. Solar panels, ground-source heating systems, wind energy and a range of heat-saving techniques are being combined to create a new generation of vernacular eco-architecture. Once again, the principles developed in the past can help us to bring beauty and sustainability into the present.

Underfloor heating

I have excavated many buildings in my archaeological career, but one of my more vivid memories is of excavating a Roman structure in Cambridgeshire in the dead of winter.

Figure 6.1: Roman hypocaust at Bath. (Photo © Ad Meskens/Wikimedia Commons)

Contrary to the stories in popular culture, professional archaeologists don't spend every minute of our working lives alight with the joy of discovery – especially not in winter. Much of the time we are simply looking to see where one layer ends and the next begins, and the change is usually quite subtle, so we need to pay attention. We are filling out recording sheets and making scale drawings – often with rain coursing over the muddy drawing film that is taped over graph paper on a large board which is heavy, awkward, and sometimes bucking about in strong winds. On one such cold, wet day I was feeling rather miserable when I made a discovery that made me stop and take stock: I uncovered a Roman box-flue tile. These tiles are not all that rare, but sometimes such finds can make you pause, step back from the technicalities, and see the past with sudden clarity. What this flue tile did was to remind me that although I was a cold, wet archaeologist in the 1990s, I was looking at a 2000-year-old tile from a building that would have housed warm, dry Roman-Britons.

What I had found was evidence for a common type of structure from the Roman period. The Romans developed an extremely durable concrete, and they built concrete floors supported by pillars made of stone or tiles, which created free-flowing airspace below the floors (Figure 6.1). Warm air from a furnace pit flowed under the floors and was then channelled up through the walls through hollow box-flue tiles such as the one I'd just found, thus warming the whole room. It is a logical way to heat a space, and underfloor heating is now making a comeback. In fact, one estate agent in Cambridgeshire had told me the previous year that it was a 'new concept' (I nodded politely). When we lived in the Netherlands many years later, we rented a flat that had underfloor heating in the bathroom, and it was heavenly. Heat rises, so it makes sense to have it emanating from the floor, not from radiators around the walls.

British archaeologists are so used to regarding underfloor heating as a Roman development that it was a surprise to me to discover that the system was actually invented independently in Korea and Northeast Manchuria, thousands of years earlier. The very earliest versions take the form of heated beds called 'kang', a word originally meaning 'to dry'.[1] A Chinese text found at Ningxia Hui and dating to 2852 BC describes heated floors or beds, but the archaeological evidence for heated flooring goes back to *c.* 5000 BC.[2] Heated floors in Korea are called 'ondol', and they were developed in around the 4th century BC. By AD 1000–1200 the fire was moved to the outside of the building so that the rooms would not overheat in the summer, so cooking and heating fires became separate. Underfloor pillars were called 'gorae', and the smoke was channelled up a chimney after it had served to heat the floors.

Ondol heating is still used in Korea, while in China it is making a comeback. In the Hubai Province some cities are putting underfloor heating into 70–80% of new buildings, and in Japan 10% more houses are built with underfloor heating every year.[3] In Korea there was at one point a concern about deforestation, and coal began to replace wood for heating after the war in 1953, but coal smoke is toxic and can leak into the rooms, so from the 1970s onwards hot water was piped under the floors instead. Attempts were made to introduce radiators, but people loved their warm floors, and ondols are now almost universal.

It's worth mentioning that central heating systems were developed during the Industrial Revolution in Britain, sometimes using ducts that introduced heat into a room at floor level. William Strutt designed a hot air furnace in 1793 to heat a mill in Derby, and during the 19th century houses for the well-to-do were increasingly built with central heating. The late 19th century house I grew up in, in Amherst, Massachusetts, had lovely wrought iron radiators, but also wrought iron grilles (we called them 'registers') placed at floor level to introduce warm air to our drafty old house. Walking through the house on a snowy winter's day, family members could be found sitting on warm registers and radiators throughout the house, absorbed in their books and oblivious to the world outside. A heavy blanket of New England snow created a silence that is palpable; a quiet stillness in which specks of dust swirl gently in shafts of winter sunlight. In many ways, that world bore a closer resemblance to the 19th century than it does to today's noisy world of ubiquitous electronics. We even had hitching posts and stone carriage steps outside the houses on my street, for long gone horses and carriages.

Underfloor heating is just one of the many practical aspects of Roman (and Victorian) architecture. The Romans didn't invent concrete, but they perfected the process, combining finely ground volcanic deposits with lime, and then adding water. One type of concrete is made with ground-up tiles, forming a distinctive pink concrete that we call 'opus signinum', which is often used to build Roman bath house floors. Durable concrete enabled the Romans to build extraordinary structures like the Pont Du Gare aqueduct in France, and the fabulous dome of the Parthenon in Rome – still standing after thousands of years. The Romans even developed a form of concrete that becomes stronger with age. Based on a mix of volcanic ash, lime and seawater, it is mixed together with volcanic rocks – and when it is exposed to the sea, *e.g.* in harbour structures, the seawater actually interacts with the concrete, causing the development of a crystalline structure which expands *within* the concrete,[4] strengthening it to such a degree that modern engineers are attempting to replicate it so that they can use it to store nuclear waste.[5]

Good concrete enabled the Romans to build strong, lasting structures, including vaults, arches and domes, but they were not the first to come up with these architectural features. A primitive form of dome structure was made using stone in prehistory, but vaults, arches and domes were also made with mud brick in ancient Mesopotamia, from around 3000 BC. Vaults are important for construction in areas with little wood for roofing, so they are an important architectural feature on the alluvial floodplains of desert regions, where trees are scarce resources. It was in such places that mud bricks were invented.

Desert architecture: mud brick

It has been estimated that between a third and half of the world's population live in houses built of earth, clay or mud brick.[6] Mud brick was developed by combining soil or sediments such as clay and sand, with water and an organic binder such as straw, reed fragments or animal dung. In Egypt, mud for mud bricks is taken from grazing land, because the animals churn their manure into the pasture soil so that it is to some degree 'ready-made'.[7] Water is added, and also straw or other organic material, which ensures that the bricks dry evenly. Dung is an important ingredient, and together with the straw it adds significantly to the cohesiveness of the brick. After the materials are thoroughly mixed – which can mean trampling by people if there are no animals about – the mixture is pressed into moulds to be shaped into bricks, which are then dried in the sun and used for construction.

The invention of mud brick arose independently in many locations. The first structures in the Levant date back to the inelegantly named 'Pre-Pottery Neolithic A' period, which began around 9700 BC.[8] Mud brick was used to build the ancient Mesopotamian cities that followed, which form the astonishingly large 'tell' sites that are so enthralling to archaeologists; as new structures were built on the ruins of the old, the towns grew higher and higher until they formed great mounds that could be mistaken for geological features on the plains. The 9 kilometre long wall around the city of Uruk is described in the epic of Gilgamesh in 2100 BC:

Go up to the wall and walk around,
Examine its foundation, inspect its brickwork thoroughly.
Is not its masonry of baked brick,
Did not the Seven Sages themselves lay out its plans?
One square mile city, one square mile palm groves, one square mile brick-pits, (and) the ... of the Ishtar Temple:
3 square miles and the... of Uruk it encloses.[9]

Ancient structures in Egypt and Nubia were also made from mud brick, and it was used in the Indus Valley in what is now Afghanistan, Pakistan and India. The Great Wall of China is partly made of mud brick, and also the American Indian Pueblo sites, which date back to around AD 700.[10] Another word for mud brick is 'adobe', from the Arabic 'tuba' (brick), which in turn derives from the Coptic word 'tobe' and the Egyptian 'dbt'.[11] The word 'adobe' travelled from North Africa into Spain, together with the introduction of Islamic architecture, and from there to the Americas.

Mud brick structures are warm in winter, cool in summer, and they absorb humidity and thus keep buildings relatively dry. They can stand up to the harsh sandstorms that whip across the desert, and they can be made from local materials by local

Figure 6.2: Mud brick tower houses in Sana'a, Yemen. (Photo © Maria Gropa, courtesy of UNESCO)

builders. They are also fire-resistant, insect-resistant, and mud brick does not rot.[12] It is the thermal properties, though, that make mud brick so well-suited to the heat of the desert; mud brick absorbs heat by day and releases it at night, keeping the temperature of the interiors fairly constant – cool in the day and warm in the cold desert nights, when temperatures in the desert can plummet.

A disadvantage of mud brick is that it needs to be regularly maintained, a process which usually means plastering the outside of the building to keep it watertight and to prevent the walls from eroding, but with regular maintenance the buildings have an extraordinary resilience. The mud brick tower houses in Yemen, for instance, predate the Islamic conquest of AD 630.[13] These extraordinary structures can be eight or nine storeys high, and before the current war there were around 14,000 in Sana'a, the capital of Yemen and a World Heritage Site (Figure 6.2). These tower houses tended to grow as families grew, with storeys being added gradually, as needed.

Another advantage to mud brick is that it is almost free of cost. Roofing material can be of wood, reeds and palm ribs, sometimes with an added layer of earth on top for extra insulation. However, where wood is scarce, the whole structure can be built of mud brick. The technology for this was almost lost, but was rediscovered by an extraordinary architect, Hassan Fathy.

Hassan Fathy and the reinvention of the mud brick vault

The Egyptian architect Hassan Fathy is one of the great visionaries who brought this ancient building technique into the modern world. Fathy graduated from the University of Cairo in 1926, at a time when concrete, glass and steel were being introduced into Egypt. Fathy resented this development, and he wanted to develop an indigenous Egyptian architecture that reflected local culture.[14] Most of all, he wanted to design buildings for the poor, in which they could live comfortably while also taking pride in their heritage. Egypt was experiencing a mood of uneasiness due to the rapidity of social and technological change, and there was a sense that traditional values were being lost.[15] Fathy wanted to address this by merging ancient tradition with modern building developments.

Fathy addressed these ideas in a book called 'Architecture for the Poor', in which he introduced what is quite a modern philosophy. He said, '[S]pace has its own logic. Islamic architecture is one of space and not walls.'[16] This is probably a statement that every country might say about its architecture, but Fathy's approach led him to study traditional Islamic architecture, with a focus on the buildings of medieval Cairo. Courtyards are an ancient feature in Egyptian architecture (as well as in wider Mediterranean architecture), and Fathy noted that internal courtyards regulate the temperature and also filter out the dust of the city, providing a quiet, private and dust-free space. Fountains were added to the courtyards in the 7th century Fustat period, which was the phase immediately following the Muslim conquest of Egypt in AD 641; these fountains added to the cooling effect, and also contributed to the sense of peace in the heart of the family dwelling. Egypt was conquered again in AD 969 by the Fatimids, and a Shia Muslim Caliphate was formed and new architectural features were developed. In this period, courtyards that formerly opened onto the streets were closed off, increasing family privacy in a city that was becoming increasingly noisy and densely populated.[17]

Fathy was interested in incorporating the architecture of these early periods, but he was also interested in the pre-Muslim architecture of the Pharaohs, including not just the impressive public monuments, but also the ordinary houses of ancient Egypt and the neighbouring Nubia. The traditional knowledge that went into creating these vernacular structures had been lost across much of Egypt, but in Nubia the techniques – which date back to at least the 13th century BC – were still in active use. Fathy went to study them first hand.

The key to building vaulted roofs, he discovered, was to begin by building a solid gable wall, against which the vault will lean. The builders marked out the height, shape and curve of the vault at the point where it would butt up against the end or gable wall. Bricks for the vaults were made with extra straw and less mud, so that they were lighter, and as they were stacked, the mud mortar was added in a wedge shape, so that it angled the bricks towards the supporting end wall. Fathy learned these techniques from the Nubian builders, and he used them in his modern designs.

Another important aspect to desert architecture is the dome, which has the practical effect of cooling the interior of a building, and the spiritual aspect of representing the dome of the sky. In simple domestic buildings, domes are made of mud brick and can be just a few metres across, but more elaborate examples can be seen in the tiled domes of some magnificent mosques (Figure 6.3). A study of stone huts with domed, corbelled roofs in the southern Levant found that they are cooler by an average of 11°C in the daytime, when outdoor temperatures are around 35°C, while remaining warm when the outside temperature drops to a night time temperature of 18°C.[18] Although less effective, roofs on mud brick houses can also be flat, and can be simply made using wooden beams covered with reeds or palm ribs or reed mats, which are then covered in mud plaster.[19] This form of architecture creates extra space, and rooftops are an important area for a number of uses, especially in crowded Middle Eastern cities. There are good ways to cool a flat roof, and experiments in Khartoum have shown that simply whitewashing a mud roof can bring down the indoor temperature by 10°C.[20]

Looking at the larger scale, traditional cities in hot climates have distinct methods of town planning that serve to minimise the oppressive daytime heat, while conserving it at night, when the temperature plummets. Streets and walkways are typically quite narrow, so that they are shaded by the adjoining walls and houses (Figure 6.4), but streets can also be roofed, as in the Turkish town of Mardin. The Greek island of Santorini has houses packed closely together in order to provide shade, with domes and vaulted roofs insulated with volcanic pumice to minimise the indoor heat (Figure 6.5). The narrow, winding streets of many desert towns also act as gentle wind tunnels, creating a cooling breeze as well as providing shade.

Figure 6.3: Interior view of domes in the Blue Mosque, Istanbul. (Photo © Shutterstock)

Figure 6.4: Narrow, shaded alley in Jerusalem. (Photo © Jerry Bond)

Figure 6.5: The village of Ora in Santorini, showing domes, vaults, flat roofs, and sheltered alleyways. (Photo © Inu/Shutterstock)

Windcatchers, badgirs and other cooling features

The windcatcher, or 'malkaf' (Figure 6.6), is a kind of hooded opening on a rooftop which catches the wind and carries the cool breeze through the house, sometimes leading it through basements where fruit and vegetables are stored.[21] Hassan Fathy used such structures when he created his new 'architecture for the poor', but he refined them by creating a series of baffles to increase the wind velocity within his structures. These Egyptian windcatchers are low forms of the Persian wind towers called badgir (windcatcher); such towers originate in the 4th millennium BC or earlier, and they are like the windcatchers but are usually much higher. They feature tall, vertical shafts with vents at the top which catch the wind and carry it down and through the buildings (Figure 6.7).

The towers are divided internally into separate shafts, so that they have an input vent for cool air and an output that is orientated in the opposite direction to the prevailing wind, so that it can suck the warm air out of the house. The partitions within the badgirs narrow as they channel the wind, which intensifies the flow. In the city of Bam, in Iran, there were badgir that were built some way away from the houses, with an underground tunnel that caused further cooling before the air was channelled through the houses.[22] In some very arid regions, mats are placed over the intake vents; these are dampened down with water, to provide a cool, refreshing breeze that blows through the house. The vents can be one directional or two, four

Figure 6.6: Rooftops in Kashan, Iran, showing windcatcher towers (badgirs) but also the low malkaf windcatchers, visible as arched openings on the central dome and flat roofs. Note also the simple, low domes in the foreground. (Photo © Aleksandar Todorovic/Shutterstock)

Figure 6.7: Wind tower or 'badgir' in Yazd Province, Iran. (Photo © Mohammad Hosseini, courtesy of UNESCO)

Figure 6.8: Mashrabiya in Old Cairo. (Photo © Khaled ElAdawy/Shutterstock)

or even eight, but the most popular were four directional. They are built of mud brick and covered in plaster for added cooling, but some are also covered in tiles, and they can be spectacularly beautiful. Some of the most impressive badgirs can be seen in the Iranian city of Yazd – a mud brick city of over a million people, now a World Heritage site because of the survival of the beautiful medieval buildings and elegant tiled medieval mosques.

Another ancient feature that Fathy recreated was the 'takhtabush', a covered seating area between two courtyards. One courtyard is a hot, paved space while the other is planted in order to create a cool, green space. The heat from the paved courtyard rises, drawing the cool air from the garden through the 'takhtabush' passageway, so people can enjoy both the greenery and the cool convection current created by the architecture.

The 'mashrabiya' is a wooden latticework screen that was originally created to prevent the women of the house from being seen by outsiders. It was discovered that having square, flat edges on the lattice causes glaring slants of light in the indoor space, but if they are rounded then it softens the indoor lighting. These beautiful features can be seen in many old buildings in the Middle East (Figures 6.8 and 6.9).

Such architectural features are not unique to the Middle East. The Chinese also have many ancient traditions in vernacular architecture that are well-adapted to high temperatures. Traditional Chinese courtyard houses are built with overhanging roofs to protect the indoors from the blazing sunlight, and latticework like the Arabic 'mashrabiya' is also used (Figure 6.10). A key feature is the light well, which is a narrow, rectangular shaft that allows air to circulate; it allows in light, but it does not allow

the heat of direct sunlight to penetrate into the building.[23] Plants such as bamboo are planted in the light wells, and screens opening onto them can be opened or shut. The layout of the house itself is such that the air moves from the front and then through a narrower area, which causes the breeze to accelerate before escaping up through the light well. Benches next to the light wells are placed so as to receive the maximum breeze, while the sitters can also enjoy the calm rustling of the bamboo.

Japanese houses were also traditionally built around several small gardens, which could be sprayed with water to set off a slight breeze. The traditional kitchen ceilings are high and contain skylights, so that all the heat from within the house rises naturally to the highest point, and escapes out the open skylight.[24]

Figure 6.9: Mashrabiya in Jerusalem, from the outside. (Photo © Jerry Bond)

Figure 6.10: Chinese house with garden, showing overhanging roofs and elaborate latticework. (Photo © Pakenee Kittipinyowat/Shutterstock)

A history of earth houses in Britain

Mud or earth houses are not just features of the desert; they are also traditional in the northern latitudes, although we do not tend to make mud bricks. In Britain, earth-built houses are called 'cob'. They do not tend to be high status houses, and some of the early descriptions can be quite snide. William Richards was an Englishman who thus described the cob houses of Wales in his 17th century satire:

> We stumbled upon a House, or a Dunghill modell'd into the Shape of a Cottage, whose outward surface was [covered] with such swarthy Plaister, that it appear'd not unlike a great Blot of Cow-turd.[25]

Richards goes on to describe the interior, where '[a] whole litter of Children was strewed upon the floor' (remember, this was satire).

Cob houses in the southern Welsh counties of Pembrokeshire and Carmarthenshire were commonly lime-washed a brilliant white, and when well-kept and regularly re-plastered they can last for hundreds of years – however, landlords often did not carry out this upkeep.[26] The Poor Law Commissioners' *Report on the Sanitary Conditions of the Labouring Poor* in 1849 describes a Labourer's house as follows:

> The cabin is so rude and uncouth that it has less the appearance of having been built than of having been suddenly thrown up from the ground... The wall... seems bedewed with a cold sweat [and] in fact crumbling to decay... The thatch is thickly encrusted with a bright green vegetation... You approach the doorway through the mud.[27]

This paints a terribly unappealing picture of earth houses, which in these descriptions sound damp, filthy and unpleasant – but they don't have to be this way. In fact, earth houses in the northern hemisphere can be healthier, warmer and drier than modern houses, even as mud brick houses in the deserts can be cooler, healthier and drier than modern concrete introductions. There are around 50,000 cob houses still in use in England, the oldest of which have been standing for 700 years.[28] Rendering the outside walls with traditional lime-wash keeps the houses looking fresh, while allowing moisture to wick to the outside where it can evaporate.

In recent studies it has been emphasised that dense earth walls hold the heat in very well, and they absorb moisture from inside the house and release it to the outside, provided that a breathable plaster is used on the exterior. This is especially important for bathrooms, where damp tends to accumulate in well-sealed buildings with waterproof rendering. Stone foundations are another important adaptation, serving to protect the walls from rising damp, while overhanging roofs protect the walls. Drier, more breathable walls also inhibit fungal growth, which is a problem in many British houses.

Our own house is a case in point. It is a stone structure built in 1886, but at some point the outside was plastered with a waterproof rendering. The problem with this is that while it keeps out the rain, it doesn't let the damp air out – a problem compounded by our double-glazed windows. Like many modernised older houses, we have damp patches that require regular cleaning, and the indoor moisture in summer is above 60% (it tends to be lower in the winter, when the central heating is on). Unfortunately, most fungal species flourish and emit spores at a relative humidity of 60% and above. If we were richer, we would strip off the airtight render on our house and either leave the stone exposed, or cover it with a breathable, traditional render. For now, we open a window (simple technology indeed) or run a dehumidifier.

It seems that in the modern quest for more energy-efficient homes, we have made a number of mistakes – all of which can be rectified, but it means a certain amount of redesign. By 2020 the UK is intending to reduce carbon dioxide emissions by 80% compared to 1990, so new houses must be air-tight and must conserve heat. The problem with this is that all the hot water that we use indoors produces steam, and we ourselves exhale water vapour; by making our buildings air-tight, we are increasing the level of dampness indoors, and the consequence is an increase in 'sick building syndrome'.[29] This isn't just an illness like the infamous Legionnaire's Disease – it is a common syndrome that causes itchy, dry or watery eyes, dry skin, stuffy or runny noses, headaches and dry throats – and it even causes a sense of lethargy and low concentration levels. If you live in a house with poor ventilation, do yourself a favour and open a window!

Part of the problem is the introduction of indoor pollutants on top of indoor damp. Fire retardants are now mandatory additives to furniture, but these compounds release formaldehyde, which is also emitted by air fresheners (a very bad idea), scented candles (who knew?), cigarettes (unsurprisingly) and some fabric softeners. In the UK, much of our women's clothing cannot be put in a clothes dryer, so we dry it outdoors where possible, or indoors when necessary – which can increase the indoor spore count to more than double what it would be otherwise. The way forward has got to be a combination of old and new construction technologies.

The future for earth houses

In 2003 it was estimated that 90% of the houses in the world are vernacular, reflecting local resources, traditions and technologies. Such buildings were regarded as 'outdated' by architects and town planners until the beginnings of a vernacular revival, which began a few decades ago but which really is still in its infancy. There is now an increasing interest in vernacular architecture, for reasons of aesthetics, cost, and sustainability of building materials, but also for practical reasons. Using local materials means that the transportation costs are low or even nil, and there is increasing interest in building new cob or earth buildings in the UK and, more widely, in Europe.

'Earth' in a European context usually means the layer below the topsoil and above the bedrock; in the UK, this is often a clayey material that was deposited by glaciers during the ice age. There are many ways of using soil, however, and in the Scottish

Highlands and Islands – and other northerly places such as Iceland – walls were traditionally built of turf. In Scotland, many older buildings have a double-walled construction, with earth packed in the middle.[30] It is tempting to suggest that this method is 5000 years old, because the walls of Neolithic villages such as Skara Brae, Orkney, are of a similar double walling – although the prehistoric houses have midden material packed between the stone walls, rather than pure soil.[31] It would be romantic to think that construction remained unchanged for 5000 years, but it is more probable that this logical mode of construction was simply reinvented over the years.

In the UK, 40–50% of our carbon emission comes from the building industry,[32] with concrete producing particularly high levels of CO_2.[33] Earth houses are an excellent alternative – there are no waste products (unlike the industrialised construction industry, which produces 50% of the UK's solid waste), and they also hold in the heat while releasing damp to the outside, and so are more efficient and save fuel costs.[34] The drier, more breathable walls are much better for stopping the growth of mildew and fungus, and so prevent sick building syndrome.

The future for earth houses is looking good. In France, the Domaine de la Terre project is a development of new earth-built houses in a region in which 90% of the traditional housing is already made with earth. So far this has proved to be a huge success: the residents love it, because the houses are beautiful, comfortable, and surprisingly modern.

Vernacular architecture and resistance to earthquakes and storms

The concept of working with nature instead of against it is important when it comes to natural disasters. Earthquakes cannot be controlled, but there are different possible approaches to building in earthquake zones. One approach is to build structures that do not collapse when shaken; another is to create buildings that cause a minimum of harm to the occupants when they collapse. Building flexible structures is a modern approach to building in earthquake zones, but it is also a very ancient one.

The vernacular architecture of Kashmir, for instance, was built to withstand both earthquakes and floods. The traditional buildings, even those of two storeys, were able to survive a major earthquake in 2005, during which the modern cement and concrete structures collapsed. There are two traditional building systems in this region: 'taq' and 'dhajji'.[35] In the 'taq' system, a dense concentration of masonry piers are built to support the floor beams. These pillars are between *c.* 75–90 cm thick, and stand between *c.* 90–120 cm apart. They stand upon a stone rubble foundation, and the inner faces of the buildings are of mud brick or rubble. Wooden tie beams separate the masonry from the superstructure, and the roofs are made of earth over birch bark overlying wooden planks, which overlie the rafters. These buildings are warm in the winter and cool in summer, as well as being cheap to build and also being very beautiful – but perhaps most importantly, they are built to withstand the earthquakes that rock this region.[36] Unfortunately, modern buildings of concrete are seen as a status symbol, and the old building techniques are being forgotten. It is to be hoped that we see a re-emergence of traditional building, or – if traditions really are to be rejected – that modern, earthquake-proof structures are built in place of the shoddy concrete buildings.

'Dajji' buildings are square, and this symmetry makes them more stable during earthquakes[37]. The walls are built of timber lattice-style frames, with brick, wood or stone infilling, which is cemented with mud and cow manure, which renders the building less brittle and allows more 'give'. The infill of one panel might fail during an earthquake, but the whole wall does not; the walls are moderately flexible, and thus are less likely to catastrophically fail.[38] The buildings are relatively lightweight, which is another advantage during seismic stress.

The traditional buildings in Nepal are also very resistant to earthquakes, and here, too we have seen the phenomenon of modern buildings collapsing in the earthquake of 2015.[39] More than 500,000 houses were destroyed in the catastrophe and 8790 people were killed in the wreckage, but many of the traditional wood, stone, clay and bamboo structures survived the earthquake. The key element is the horizontal and vertical tie beams of wood or bamboo that hold traditional buildings together. Wood being expensive, the use of steel tie beams has been proposed as a marrying up of old and new building traditions, or alternatively the introduction of 'seismic bands' of polypropylene mesh or even concrete, which would serve to hold the buildings together. Currently, mud brick and stone houses are being retro-fitted with prefabricated bamboo mesh structures both inside and out. The bamboo staves are set 10 cm apart, and are held in place by heavy gabion wire. The cost is minimal, and the added structures can then be plastered over with mud so as to be invisible.

The vernacular architecture of Erzurum has also been held up as an example of earthquake-proof traditional housing. Erzurum is a city in the east Anatolia region of Turkey, and in the 14th–17th centuries the city was renowned as a garden city. In the early 19th century, houses were embedded in the ground and had grass roofs where animals grazed, but following an earthquake in 1859, in which 65% of the city's housing was damaged, a new tradition emerged.[40] After 1859, houses were built above ground and made use of wooden beams that made the structures very resistant to earthquakes. These buildings – now regarded as 'traditional' – were made of thick walls with a rubble core, which kept the houses warm in the cold, snowy winters of the region. Modern architecture in the region is now largely built of concrete, which is thermally inefficient, and it is also proving to be more susceptible to earthquake damage.

Another particularly earthquake-proof style of housing is the 'Paisa House', seen in the central mountain region of Colombia. These combined houses/industrial buildings are made of bamboo, wood and adobe, and they are exceptionally good at withstanding earthquakes. They are built into the mountain sides, and are used for drying coffee beans as well as for accommodation for the coffee pickers. The flexibility of the materials makes them extremely resilient.

The traditional houses of the Chittagong Hills in Bangladesh are also well-adapted to both hillslopes and earthquakes. The houses are built on bamboo stilts, which create rows of pillars that distribute the weight of the roof. The pillars are propped up on three sides, and all elements of the building are lashed together with split bamboo, rather than nails. These sturdy structures withstand strong winds and the heavy storms of the monsoon season; one observer noted that they also stand up to large groups of people dancing and jumping up and down. In 1869, another observer noted,

> A hill house perched in an exposed position on the ridge or spur of a lofty eminence looks the frailest structure in the world; its strength however is surprising, and in spite of the fearful tempests that sometimes sweep over the hills, I never heard of a house having fallen or being injured by the wind.[41]

Bamboo and wood are key to the styles of structure discussed so far, but another approach to creating storm-proof buildings is to build a solid central structure with more ephemeral walls or roofs. On Santorini, the traditional houses have thick walls built with thin, vaulted, buttressed roofs. Pumice covers the vaults, creating a lightweight finish to the roofs which also helps keep the buildings cool in summer, while at the same time creating a flat, usable roof space – and the additional benefit is that these lightweight roofs and strong walls are good at withstanding earthquakes.

My final example is from Samoa, where the traditional Fale Tele houses are round, with wooden poles in the centre and a ring of roof supports around the outside. The sides are left open, but can be closed with blinds made from woven palm leaves. When Samoa was hit by massive cyclones in 1990 and 1991, the traditional Fale Tele houses survived much better than concrete housing, and most importantly the lightweight walls could not fall and crush the inhabitants.

Floating houses

Both developing and developed countries are experiencing increasing numbers of floods, which are increasingly catastrophic in their effects; in the last 20 years, the ten worst floods alone have killed 500,000 people and displaced many more.[42] As the global climate warms, rising air and sea temperatures and increased water vapour are leading to more frequent, more severe storms – as well as rising sea levels. There are several architectural responses to such conditions, and when we look to the past we can see two age-old methods of living with water: pile dwellings, in which structures are raised above the level of the highest local water level, and floating houses. In the past, pile dwellings and floating homes were often built for safety, in order to place houses and villagers out of reach of attackers. They were also used in order to adapt to changing water levels, for instance in Vietnam's Halong Bay, where floating houses were built in response to rising sea level (Figure 6.11). A similar approach was developed on the Tonlé Sap, a lake in Cambodia with water levels that fluctuate by

Figure 6.11: Floating houses at Halong Bay, Vietnam. (Photo © Elizabeth Winterbourne/Shutterstock)

up to 9 m; here, the locals build a combination of houses on high stilts in some places (Figure 6.12), and also floating houses similar to the ones in Halong Bay. This is a way of dealing with fluctuating water tables, but it is also a response to the shortage of land for building – a problem afflicting many cities around the world. One response to urban land shortage has been to build floating neighbourhoods, for instance in Seattle, Vancouver and Hong Kong, and new floating villages are being developed in the Netherlands and the UK.[43]

In the Netherlands, the 'Room for the River' programme set aside large areas of land for floodplain, where it is illegal to build conventional houses because of the new regime of periodic controlled flooding.[44] One of these areas is along the Maas River at Maasbommel, where a floodplain and dykes protect the old village. On the floodplain on the river side of the dyke, a cluster of new 'amphibious houses' have been built, together with 14 fully floating houses. Both have been built on hollow concrete cores that float on water, and they are held in place by large piles which allow the buildings to move vertically up and down by about 4 m.[45] Services such as water and electricity supply are protected in flexible, waterproof pipes so that they also rise and fall with the water level. These houses have been a great success, but they are only the beginning; there are now plans for many more such floating villages in the Netherlands and other regions where floods are common or potentially catastrophic, such as New Orleans.

Figure 6.12: Houses on piles at Tonlé Sap. (Photo © DeltaOFF/Shutterstock)

Figure 6.13: Amphibious House in the Thames. (Photo © Oliver Pohlmann, courtesy of Baca Architects)

Figure 6.14: Amphibious House in the Thames, back of the house. (Photo © Oliver Pohlmann, courtesy of Baca Architects)

One particularly fine example of an amphibious house is on the Thames, where there has recently been catastrophic flooding. There, a house has been designed to float up and down within a concrete holding tank, with a range of 2.5 meters (Figures 6.13 and 6.14).[46] This is well within predicted flood levels, and tests show the house is well adapted to its watery surroundings and should be future-proof. The company that designed this house, Baca Architects, specialises in the new field of 'aquatecture', in which buildings are designed to function in an amphibious or watery environment. This new field is going to see some exciting new developments, particularly as the Royal Institute of British Architects has issued a statement emphasising that it is no longer good enough to simply try to prevent flooding – we have to start finding better ways to build in flood-prone areas.[47]

Architectural anthropology

Architectural anthropology is a new field of study that investigates how people in different cultures use buildings and space. Planners from Europe have a long tradition of trying and failing to help developing countries, because they fail to understand local needs. Hassan Fathy, the Egyptian architect, designed an entire village – New Gourna – with the intention of creating a beautiful and traditional space for the old village of Gourna, which the government wished to relocate so that it was a good distance away from an important archaeological site that the villagers were systematically robbing. Unfortunately, the new village was not a success, because the people had no way to make their living in the new location, but the concept seemed like a good one. Instead of mass-producing identical buildings, Fathy ensured that each was distinct. 'In nature,' he said, 'no two men are alike. Even if they are twins and physically identical, they will differ in their dreams. The architecture of the house emerges from the dream. That is why, in villages built by their inhabitants, we will find no two houses identical.'[48]

One of Fathy's more controversial ideas was to build the village without running water – but there was a reason for this. Collecting water from communal wells was regarded as a woman's job, and in the traditional Arabic culture this was the only time a young woman could leave the house. It was a chance to get out and to talk, but also it was the only time that men could see them, and so this was partly how women found husbands – the men would gather to watch the girls.[49] As we have seen, many of the features of traditional Islamic architecture were created in order to keep women hidden, and Fathy wished to keep this one feature that would allow women to get out into a public space (although having read his remarks, it seemed that his concern was largely for the young men who wished to see the girls, rather than the girls' wish to get out of the house).

In Turkey, the desegregation of women is changing the way houses are laid out. Traditional buildings had men's and women's areas, but the layout is changing because men and women no longer wish to lead such separate lives.[50] Marriages, it seems, now involve a greater degree of friendship and companionship between husband and wife, and buildings are being designed accordingly.

Planning for natural disasters has shown us another aspect of human nature and culture. Disaster relief workers used to assume that the newly homeless needed first of all to be protected from the elements, but in practice this is not always the first priority of the disaster victims. People often prefer to camp on the site of their ruined home, because it provides a sense of having a 'territory', which provides emotional security and a sense of having a place in the world. The family address may also be necessary in order to receive food and medical assistance, and people who have lost their homes will often still need to commute to work.[51] Disaster housing also needs to include a water source – a rather surprising omission in some poorly planned instances – and it needs to take family size into consideration. Western aid agencies have been known to provide disaster accommodation for nuclear families, which are inappropriate for cultures where people live in extended families.

One interesting instance in which an aid agency got the culture wrong was an African project for refugees, in which 300 houses were built. They were beautifully constructed with domed roofs – and unfortunately the style was exactly that of the distinctive funerary structures in which the refugee population traditionally laid their dead to rest. No one was prepared to actually live in these houses, preferring tents to mausoleums.[52]

Urban farms and gardens

In the towns and cities of later Roman Britain, there is a distinctive type of layer called 'dark earth'. This is not the same as the Amazonian Dark Earth, but rather it is a dark deposit that is high in phosphate and contains lots of small fragments of building rubble. Initially these layers were thought to be wind or water-lain, but they lack the 'sorting' that characterises these sediments (*i.e.* they aren't pure sand, silt, clay or gravel, but are very mixed).[53] They occur as early as the 2nd century AD, and sometimes have Roman structures built on top of them, but they are mainly from the 4th century or later. They tend to be high in phosphates, which are derived from human and animal excrement, and they often look quite homogenous.

There have been many different interpretations for these soils, and it now seems reasonable to conclude that more than one of the interpretations are correct, because dark earth soils are actually more variable in content and derivation than was first supposed.[54] Some are formed by distinct dumps of rubbish, while others have collapsed walls or timbers poking up through the layers, and so simply represent vacant lots.[55] Animal manure and human excrement ('nightsoil') were dumped in these vacant lots, but this form of waste-disposal also served to fertilise the land, and it appears that some of these increasingly fertile spaces became gardens.

It seems the Roman towns were shrinking, becoming depopulated, and partly falling into ruin – but another way to look at this was that people were simply living a different kind of urban life than in the earlier centuries. As buildings collapsed and land fell out of use, sediment and waste built up, and people seized the opportunity to create vegetable plots and orchards in the vacant lots.[56] In the 4th century we start to find agricultural buildings in Romano-British towns, including granaries and corn drying kilns, which suggest a form of urban agriculture.[57] This interpretation is supported by the fact that some of the sites that are interpreted as gardens in the

Late Roman and early Saxon periods continued to be used as such right up into the medieval and later periods.[58] At this point, we have historical records that confirm that these sites were gardens.

Many years ago, I directed an excavation in advance of a new supermarket in the centre of the small town of Berkhamsted. The excavation uncovered the back garden plots of a row of 12th–13th century houses, and the environmental evidence was the most interesting aspect of the dig. The pollen indicated that there were orchards nearby, and we found horseshoe nails and evidence for stable floors. These back gardens would have had chicken coops, vegetable gardens, stables, and possibly a pig or two. As I was doing some research in the local library, I came across a real estate map linked to the sale of one of the houses in the 19th century, very near to the site on which I was working. The map showed, in a beautiful coloured pencil drawing, the river that backed onto the gardens, and also orchards, stables, pig pens, chicken coops, a manure heap and an orchard. In short: these large gardens were used in much the same way from the 12th century all the way up to the 19th. Town and country were often not as separate in the past as they are today, with towns being more sparsely populated and town dwellers being more self-sufficient than today's urbanites.

Gardens in towns make sense, for a whole host of reasons. In the Middle Ages, food could not be transported long distances, and most cities were surrounded by a green belt of farms that produced the food for the city dwellers. At night, manure and human excrement from the towns was loaded onto horse-drawn carts and taken out to the countryside to fertilise the fields (this is why human excrement is sometimes called 'nightsoil'). Farms could also be located within cities and towns. In Paris, for instance, around 1400 hectares – which was around one sixth of the city – were used for urban farms, right up until 1918.[59] Before the advent of cars, the horses that pulled the carts, carriages and buses in Paris produced about a million tons of manure per year. This was collected from streets and stables and delivered to farms within the city, providing fertiliser so that the city farms could provide Parisians with local food, while at the same time disposing of the waste.[60] The creation of such large, commercial market gardens is one approach to feeding the city, while individual allotment gardens are another. In Britain, allotments were rented out privately to town dwellers up until 1887, when new legislation was introduced that required councils to provide them for the populace.[61] City planners in 19th century Britain also recognised the importance of allotments for the urban working class, who did not have access to gardens, and who were often country people who had come to work in the new factories.

The notion of 'garden cities' was further developed in 1898, when Ebenezer Howard published his *Garden Cities of Tomorrow*. His idea was to build new towns in which 5/6 of the area would be used for growing food, with large gardens and allotments all around the city. He aimed to make British cities self-sufficient, and while the idea didn't catch on immediately, it was revived in the 20th century.

The need for British self-sufficiency became urgent during the First World War. The number of allotments tripled between 1913 and 1917, and although the use of allotments diminished after the war, they became popular again during the Great Depression.[62] The idea was revived again in the Dig for Victory campaign of 1939, which led not only to the wider use of allotments, but also to people digging up their flower beds to plant vegetables. Even public parks were dug up and planted with cabbages, and by 1944, allotments, gardens and public parks together were providing 50% of the nation's fruit and vegetables.[63]

The use of urban space for growing food has waxed and waned, depending on a variety of factors. The end of long distance trade networks at the end of the Roman Period, coupled with a drop in population, led to more agriculture in the cities. Wartime shortages, economic slumps and urban poverty have been factors that led to the modern revival of urban gardening. It seems sensible to me that city planners should factor in space for individual and community gardens, not only as an appealing amenity, but also as a practical means of building independence and resilience for our city dwellers – and while it is to be hoped that there will be no economic or military upheavals in our future, it makes sense to be prepared.

Cuba: necessity as the mother of reinvention

Cuba is a remarkable example of a modern country that had to undergo a sudden and rapid return to using traditional knowledge and to urban gardening, because this small island nation was quite suddenly disconnected from the world of abundant fossil fuel. The collapse of the USSR in the 1990s led to the near-collapse of the Cuban economy, because 80% of Cuban trade had been with the USSR and the Eastern Bloc countries. Cuba had been importing all of its wheat, 48% of its fertiliser and 82% of its pesticides.[64] Quite suddenly, the trade structure collapsed, and the Cuban government had to act quickly. They chose to decentralise their agricultural system in order to produce more food.

In a departure from top-down dictats, the government sought the opinions of the citizens, and particularly the opinions of older farmers who had experience with pre-industrial agriculture.[65] Fidel Castro admitted that the large scale, state-run farms had not been very successful, and so the State farms were decentralised and devolved into smaller co-ops.[66] Farm workers were given a share of the profits from all crops that were surplus to the production plan, and they were also given plots on which they could grow their own fruit and vegetables.

The Cuban government also launched a kind of 'dig for victory' campaign, in which they encouraged city people to grow more of their own food.[67] Havana, the largest city, had begun a self-sufficiency drive back in the 1960s, when the government established a green belt around the city in order to provide food for the capital. The green belt was revitalised and expanded in 1991, when Havana launched a campaign that encouraged the people to use every available space in and around the city for growing food. Urban gardens sprung up everywhere, initially as a spontaneous approach to food shortages, but later with the help and support of the government.

All undeveloped urban land was given over to agriculture, free from taxes and with the help of the State.[68]

The new urban gardens allowed vegetable production to double between 1994 and 2000.[69] By 1999, urban agriculture was providing 50% of city people's daily vegetables,[70] rising to 70% by the year 2000.[71] The city composted organic waste on a large scale, which was helpful to urbanites who had limited space, and the rich compost that was produced increased the productiveness of the urban plots. Raised vegetable beds were created out of any material available, including building rubble, stones, loose bricks and rubbish. The resulting beds are consequently scrappy and irregular, and while they are not as ornate as they could be, they are cost effective. They have also grown and developed over time, often becoming local amenities. Urban farms have been set up in cooperation with schools, medical centres and old people's homes, providing outdoor space for socialising and child minding, as well as space for gardening.

There has been an improvement in the air quality in Havana, because of the increased number of trees, and there has also been an enrichment of the soils. The government's policy of supporting organic agriculture has cleaned up both the urban and rural environments, and reduced tillage is enabling the soils to hold on to more carbon. What began as an economic disaster has gone on to show the world what can be done when sustainable agriculture is implemented in an industrial nation. While none of us would like to experience the sudden shock of a market collapse, what Cuba shows us is that it is possible for an industrial agricultural economy to turn very rapidly to sustainable and largely organic methods. We do not have to choose, in the words of Earl Butz, 'which 50 million [people] we are going to let starve or go hungry'.

Notes

1   Bean *et al.* 2010.
2   Bean *et al.* 2010.
3   Bean *et al.* 2010.
4   Ahmad 2017.
5   Jackson *et al.* 2017.
6   Little and Morton 2001, 7; May 2010, 24.
7   Dabaieh 2013.
8   Kuijt and Goring-Morris 2002.
9   Postgate 1992, 74.
10   Fazio *et al.* 2009, 257.
11   Dabaieh 2013.
12   Dabaieh 2002.
13   May 2010, 78.
14   Steele 1997, 9.
15   Steele 1997, 9.
16   Steele 1997, 11.
17   Steele 1997, 11.
18   Oliver 1987, 114.
19   Emery 2011.

20　Oliver 1987, 124.
21　Steele 1997, 15.
22　Ghaemmaghami and Mahmoudi 2005.
23　Lau *et al.* 2014.
24　May 2010, 88.
25　Richards 1682, cited in William 2010.
26　Little and Morton 2001.
27　Mackay 1849, cited in William 2010.
28　May 2010, 54.
29　McGill *et al.* 2015.
30　Little and Morton 2001.
31　Ritchie 1995, 29–30; Simpson *et al.* 2006.
32　Little and Morton 2001.
33　May 2010, 43.
34　Little and Morton 2001.
35　Beg 2016.
36　Beg 2016.
37　Shah and Tayyibji 2008.
38　Shah and Tayyibji 2008.
39　Adhikary 2016.
40　Kirbaş and Hizli 2016.
41　Lewin 1869, cited in Rashid and Ara 2015.
42　Barker and Coutts 2016, 18.
43　Barker and Coutts 2016.
44　www.ruimtevoorderivier.nl/english/ Accessed 9 November 2018.
45　http://news.bbc.co.uk/1/hi/world/europe/6405359.stm Accessed 9 November 2018.
46　https://www.dezeen.com/2016/01/20/baca-architects-bouyant-amphibious-house-river-thames-buckinghamshire-floating-architecture/ Accessed 9 November 2018.
47　https://www.architecture.com/-/media/files/press-release/riba-flooding-policy-paper.pdf?la=en Accessed 9 November 2018.
48　Steele 1997, 192.
49　Steele 1997, 64.
50　Kirbaş and Hizli 2016.
51　Davis, in Asquith and Valliga 2006.
52　Özkan 2006.
53　Rogers 2011.
54　Speed 2014, 117.
55　Speed 2014, 58.
56　Macphail 1981.
57　Rogers 2011.
58　Macphail 1981.
59　Girardet 1999, 53.
60　Girardet 1999, 53.
61　Howe *et al.* 2008.
62　Howe *et al.* 2008.
63　Howe *et al.* 2008.
64　Díaz and Harris 2008.
65　Díaz and Harris 2008.
66　Wright 2009, 138.

67 Díaz and Harris 2008.
68 Wright 2009.
69 Viljoen and Howe 2008.
70 Henn and Henning 2002.
71 Wright 2009, 90.

7

The Tao of Environmental Management

Returning is the motion of the Tao.

Yielding is the way of the Tao.

Lao Tsu: the Tao Te Ching

In this book I have aimed to focus on the success stories of the past, but in doing so it has also been necessary to discuss some of our failures. Figure 7.1 is a pictorial summary of some of the mistakes we have made, and are making, in our landscape management. Bear in mind that our ancestors first created agricultural terraces thousands of years ago, and remember also that terraces have been reinvented independently at different times and places ever since – and yet, this egregious destruction of our vital soil is nevertheless taking place all over the world today. Forests are being clear-cut from steep hill slopes, causing tons of soil to erode down slope and into the lakes and rivers on the valley bottoms. Not only is the precious soil lost, but the sediment pouring into our watercourses causes lakes and river beds to become shallower, which cause rivers to burst their banks and wreak destructive havoc on surrounding towns and farmland. Furthermore, when lakes become shallower they also become warmer, which means that fish and other aquatic species can die out. The maddening thing is that there is no *need* for this senseless waste.

The consequences of poor management
Contrast this desolation with the extraordinary care with which the Chinese maintain their rice-paddy-fishponds and terraces; contrast this with the complex, man-made agricultural ecosystems that we see all over the developing world, wherever traditional knowledge of sustainable agriculture has not been lost.

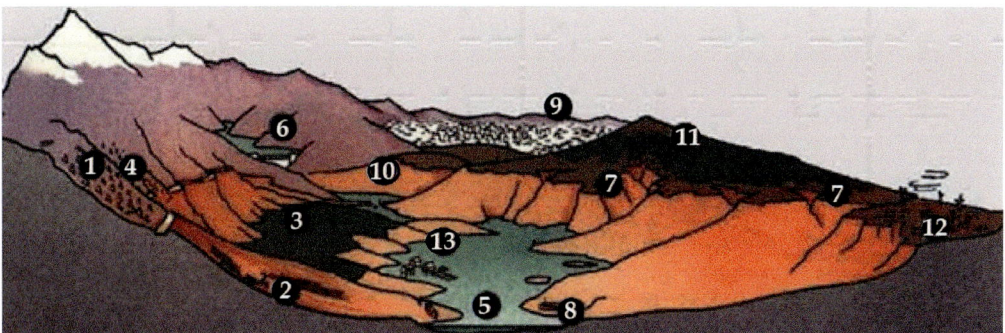

1. Deforestation
2. Steep land being cultivated down the slope
3. Monocrops grown over large areas
4. Landslide blocks road 5. Fish catch reduced in shallow waters
6. Siltation cuts hydroelectric plant's lifespan
7. Gully erosion eats into crop land
8. Mud banks reduce navigability of rivers
9. Urban slums grow as rural population migrates to the city
10. Bridge destroyed by floods
11. Crops grown on large unprotected fields
12. Wind erosion affects badly managed pasture
13. Frequently flooded village is deserted

Figure 7.1: Eroded dystopia. (Reproduced with permission from the Food and Agriculture Organization of the United Nations (FAO 1995))

Is it too late to go back to farming on agricultural terraces? Of course not! Agricultural terraces are still used throughout Europe in the old wine growing regions, and terraced vineyards have been listed as World Heritage sites in Portugal, Italy, Switzerland, Germany and Hungary. Terraced rice paddies are widely used and maintained all across Asia, particularly in China and the Philippines. Ancient terraces on Mount Carmel in Israel have been rebuilt, in order to grow sustainable forests and orchards.[1] Inca terraces are being rebuilt and cultivated in Peru, where water is a scarce resource; archaeologists had found that terrace soils hold moisture even six months after they were last irrigated, and the revitalisation project is so successful that the idea is spreading now to other regions in South America.[2] Crumbling medieval terraces have been rebuilt in Yemen,[3] while in Europe new terraces are being built to support new vineyards.[4]

We can also improve our soil conservation practices in other ways, such as ploughing with the contour instead of up and down. Planting trees on the brow of hills can be done anywhere, in developed and developing countries, and low-tillage or no-tillage agriculture is so much more cost effective that the practice is now spreading rapidly around the world. Reforestation in Africa is providing shelter and firewood while reducing wind erosion, and where erosion is reduced, waterways

1. Reforested land
2. Gully erosion halted by check dams and trees planted on gully banks
3. Steep land is bench-terraced
4. Contour cultivation practiced on lower land
5. Bunds are built to control surface runoff
6. New reservoir supplies power to nearby villages
7. Shelter belts reduce wind erosion, pastures are improved or upgraded
8. Crop rotation practiced in strips along contours
9. Tree crops grown on eyebrow terraces on steep land
10. Forested slopes prevent siltation of reservoirs

Figure 7.2: A sustainable landscape. (Reproduced with permission from the Food and Agriculture Organization of the United Nations (FAO 1995))

become cleaner and more able to support fish stocks. Each small step in promoting soil conservation has many positive knock-on effects.

While Figure 7.1 is a grim reminder of the appalling loss of land and soil that is taking place in many countries, Figure 7.2 illustrates that we have the technology to change this dystopian landscape into a green and sustainable vision for the future. In this idealised landscape, terraces and contour ploughing help to prevent soil erosion, while forests at the top of the hills prevent floodwater from causing erosion lower down the slopes (a well-established technique in much of Asia). Re-foresting the hilltops can also diminish or prevent floods from occurring further down in the watershed, because of the quantity of water that is taken up by the trees. These techniques could also be supplemented by creating smaller fields – a prehistoric land management technique that also saves the soil. Replacing old hedges creates biodiversity, cleans our air, saves our soil, and harbours insects that eat crop pests.

Saving water

One of the key things we can learn from the environmental successes and failures of the past is that we need to find ways of working with nature, not against it, and we need to conserve the resources on which we depend. Here in Europe, we are beginning to change the way that we govern and engineer our environment, in order to adopt this approach of 'going with the flow'. One example is in the new ways we are finding to manage water, which in Europe is causing increasingly catastrophic flooding. Equally, water can be in short supply in summer, even in Britain, where we occasionally have heat waves that lead to bans on watering our gardens. Both shortages and floods result from poor management.

The ancient Middle Eastern systems of underground canals and water cisterns were an efficient way to keep water from evaporating. Our modern reservoirs, by contrast, are efficient in collecting water but terrible for storing it – until now. An Israeli company has developed a cheap and flexible new solution that reduces evaporation in reservoirs by up to 94%, while at the same time cooling the water temperatures by up to 6°C.[5] The system involves covering the reservoir surface with floating balls that rise and fall with the water level, covering the water surface and thereby reducing evaporation. The balls take in water, which condenses inside the ball and then drips back out again, so that each individual ball also contributes a cooling effect to the water. The continual movement of water in the balls ensures there is no standing water to give rise to algal growth within the balls, and they also maintain the water quality of the reservoir by preventing algae and bacteria from proliferating in the reservoir itself. They are good for fish farms as well as reservoirs, because they prevent the fish from being eaten by birds. This inexpensive new system is expected to double Israel's water supply, and it is expected that its use will take off in California and other arid regions – particularly in Africa, where Israel has various water aid programs.

Another new development is a portable solar powered desalinisation plant, which can process up to 10,000 cubic meters of water per day.[6] The system obviates the need for electricity altogether, because it uses heat from the sun to evaporate the water, and the water vapour is then cooled and collected. This system cuts the cost of desalinisation by 90%, and it is attracting interest in Africa, Asia and America. Coupled with drip irrigation, this will be another step towards providing water to people and to farms in deserts in coastal regions all over the world.

Rainwater collection and storage, once regarded as an outdated method of water provision, is increasingly being adopted in the developed and developing world. Collection from rooftops is taking place in Singapore, Thailand, Japan and in a number of countries in Africa, and also in Berlin and Hawaii – just as a few examples. In India, a law was passed in 2001 that made it mandatory for new buildings with large roof areas (over 100 m²) to install rainwater collection systems. The law also applies to areas of new development on plots of land over 1000 m². Other arid countries are also introducing new rooftop rainwater catchment systems. In Jordan, where rainwater harvesting was once the norm, it is now making a comeback, especially since the government made it a requirement to build water collection and storage tanks on all new homes.[7] Some engineers, especially in the Middle East, are looking at restoring ancient practices of rainwater harvesting for drinking water, but with new safety features incorporated into the systems. A recent experiment in Egypt, for instance, found that water collected from rooftops can be filtered through a sand filter and then passed through a UV lamp for disinfection, making it safe to drink.[8]

Rainwater harvesting is also being reintroduced for use in agriculture, inspired by the ancient farms of the Negev and wider Levant. New research is taking place into the best places to set up these new farms, and these new studies are using satellite and climate data together with spatial modelling for maximum efficiency. The new research is considering things like soil properties, topography, land cover, land use and surface runoff potential, and is using hydrologic modelling and remote sensing incorporated into GIS models.[9] Jordan is one of the main developers of this new technology, but it is also taking place in other Middle Eastern countries, the Philippines, Brazil and the countries of North Africa. When I first proposed this kind of merging of old and new technology, people thought I had completely lost the plot. Now, it is recognised as common sense.

Revitalising the countryside

I aim to dispel the notion that sustainable agriculture is a costly or inefficient use of farmland. I am not suggesting that we go back in time, or that we reject science, but rather I am suggesting that traditional practices can be adapted to modern agriculture in ways that are beneficial to the environment, *and* that are cost effective. Farmers around the world have demonstrated the many economic and environmental advantages of sustainable agriculture. A recent study in Germany showed that wheat, oats and rye yields increased after farmers went organic, largely because of the vastly increased quantity of organic material in the soils.[10] In Switzerland, sustainable farming is so successful that the government is paying people to switch, because helping with the initial costs will help to nudge farmers in a direction that will help them in the long run: the farmers will benefit financially, the environment is being revitalised, soil is conserved and everybody wins[11] (except Monsanto). In Honduras and Guatemala at least 45,000 farmers have managed to triple their maize production by switching to sustainable methods, and the consequences have been beneficial not only to the land, but also to the society itself: the new, sustainable systems have created many local jobs, which have led to young people returning from the cities.[12]

There is currently an interest in renewing old breeds of farm animal and traditional varieties of vegetables and grains. There are many reasons for keeping old varieties alive, including their tolerance of extreme weather conditions and resistance to crop diseases. We know that there is security in biodiversity, and yet we are engaging in some risky practices. For example, it has been discovered that the entire soybean industry in America is derived from just six species.[13] This lack of diversity means that these soybean crops – and many other crops with a restricted gene pool – all have the same vulnerabilities to disease, pests, or environmental extremes. We need a return to diversity, to growing a range of crops and raising a broad range of domestic animal species which have a range of different characteristics and resistances. This would raise our food security, and there are also some health benefits. Some of our old crop varieties, for instance, can be tolerated by people with allergies to modern hybrids, particularly wheat.

In a revitalised countryside there are jobs for those who wish to remain rural, but what about those who would prefer to live in the city? What about the young people who want to go to urban universities, or to enjoy the buzz of city life? In cities, too, we can learn from the past, while at the same time incorporating new technologies that make the urban environment healthier, greener, more sociable and connected and also less of a drain on the planet's resources.

Sustainable urbanism for greener cities

Soft engineering

Modern cities are often poorly designed in terms of water management. Rather than collecting and storing rainwater, 19th and 20th century planners set up drainage systems in order to get rid of the water that drains off roofs, roads, car parks and other areas of concrete. Instead of percolating into the ground and being absorbed by trees or standing in wetlands, rainwater in built-up areas flows off the paving and into the drains. It then races from the drains into the rivers, creating storm surges which often cause rivers to burst their banks.

There is now a new approach to this problem: the new 'soft engineering' aims to work with water rather than simply trying to get rid of it as quickly as possible. The new engineering includes rainwater storage tanks on rooftops and in reservoirs in city squares. Water stored on rooftops can be used to flush toilets or for roof gardens, while water in ground-level reservoirs can be used to irrigate municipal parks. Architects are also getting involved, and in the new field of 'aquatecture', architects and planners are looking at new ways that people can live with water. The ideas they are coming up with include building houses on stilts and building amphibious houses that float when the water rises, as well as building houses on high ground like the ancient terps, and building houseboats. Flood relief parks are being built, which are green spaces and playgrounds when it is dry, and lakes when there are floods.[14] Village greens become 'village green/blues', readily transforming from open land to lakes and back to dry land, but with yet another advantage built in: the open space can also be used for ground source heat pumps. Gardens are built in channels that turn into flood relief canals when the water rises. Architects are thinking big, and are considering each part of the river catchment instead of looking at just one house or neighbourhood on its own. Two inspiring architects, Robert Barker and Richard Coutts, describe the new thinking as follows:

> 'Upper catchment: let rain slow'. In the hills, woodland can be planted to take up excess water, so that heavy storms have less of an impact on rivers.

'*Middle catchment: let river flow*'. Here, water is channelled so that it keeps to its banks. (There are also moves to reinstate floodplains, so that water has somewhere safe to go, rather than flooding our communities.)

'*Lower catchment: let tides go*'. Where the rivers meet the sea, allow the water to move freely. High tides and surges are allowed to flow around development areas, with houses floating up and down with the changing water level, or safely on piles up above the flow of the tide. The plans for these tidal environments include micro-hydroelectric stations in which Archimedes Screws are turned by the tide.[15]

Porous paving is another new 'soft engineering' approach to dealing with excess rainwater. Porous or permeable paving allows water to percolate *through* paving rather than running off it. This can go a long way towards mitigating storm surges that can cause flooding during heavy or prolonged downpours. Permeable paving is usually laid down as part of a sequence of layers, including layers of rock and soil, but the system can also include geotextiles – durable materials that filter out sediment, organic material and even pollutants. Adding nutrients to the geotextiles before laying them down enables them to support microbes that break down organic detritus and oil, so the system can also be self-cleaning.[16] Porous paving alone removes up to 99.6% of the oil from runoff water, and so prevents it from entering streams and groundwater, but with added geotextiles the filtering, de-toxifying effect is particularly efficient.[17]

Green roofs and green walls

Green roofs are another element of the new engineering. They can absorb up to 80% of the rain falling on a building, but they do more than that: they also keep buildings warm in winter and cool in the summer. A study on a green roof in southern California, for example, found that the roof cooled the building so effectively that it reduced the energy consumption used by air conditioning by 75%.[18] Green walls are a similar concept: one superstructure is built for soil and another is built to provide a trellis for the vegetation to cling to, so it does not damage the building itself. The vegetation becomes like cladding, providing an attractive exterior that also cools the building in summer and insulates it in winter (Figure 7.3).

The notion is not entirely new; building turf structures and turf roofs is a very old approach to housing in northerly latitudes where wood is scarce, and turf roofs are known for their insulation properties. Turf roofs may date back to the Neolithic in Britain, and were commonly used in the Viking period.[19] They were used right up into the 19th century in Germany and Scandinavia,[20] and were used throughout the North Atlantic, for instance in Iceland, Greenland, and on the traditional roofs of the Hebridean blackhouses of Scotland (Figure 7.4). It may seem counter intuitive, but a cover of soil and vegetation actually preserves the roof structure of both flat and sloping roofs, because the soil protects the roof from the damaging effects of weathering and ultraviolet light.[21] The soil does not actually rest on the roof, but rather on one or more 'vegetation support courses'. These layers can be of shale,

Figure 7.3: Green wall. (Photo © Evannovostro/Shutterstock)

Figure 7.4: Turf buildings at the Glaumbaer Museum in Iceland. These date to the 18th and 19th centuries, but are similar to the structures of the Vikings who settled Iceland in the 9th century. (Photo © Roman Shatkhin/Shutterstock)

pumice, lava, terra cotta, clay, slate, brick or other materials. A key element is the remarkably durable roof membrane, which is impervious to both roots and water.

The actual vegetation on a green roof is dependent on local conditions, but ranges from simple, low-maintenance sedum roofs to elaborate roof gardens which can include trees, benches, gardens and water features. Roof gardens on schools, apartment buildings and community centres can be used to grow vegetables, and to provide a green space for relaxation.

The benefits of green roofs go way beyond insulation: first of all, they absorb water, so they lessen the impact of storm water that would otherwise create storm surges. They also reduce summer heat, not just inside the building but also in the surrounding area; a recent study of buildings with both green walls and roofs found that the air temperature around the buildings was reduced by up to 11°C.[22] This enables people to turn down or dispense with air conditioning, thus saving fuel and reducing carbon emissions. Yet another benefit is that green roofs even out the temperature on the rooftop, which enables rooftop solar panels to operate more efficiently.[23]

Green roofs also absorb the airborne particles that contribute to urban heat, and they absorb pollutants and carbon dioxide – while at the same time emitting oxygen. The air around green-roofed buildings is cleaner, and when you add green walls, the fresh, clean air is really noticeable to pedestrians at street level. There is another effect that is worth noting: to add to the calming effect of greenery and fresh air, green roofs and especially green walls have been shown to absorb the noise of the city.[24] They reduce reverberations and dampen loud noises, and so add some tranquillity to the urban experience.

Green roofs are not a costly luxury, but are a practical solution to a myriad of problems, being both cost effective and adding peace, beauty and clean air to our towns and cities. They are more expensive to install, but they prolong the life of roof structures, paying for themselves and reducing the running costs for the tenants or businesses that they house.[25] We are increasingly seeing green roofs on airports, which typically have large expanses of flat rooftops, and green roofs are increasingly becoming an aspect of urban planning. A quarter of the flat-roofed buildings in Stuttgart now have green roofs,[26] for example, and in Tokyo the Metropolitan Government requires all developments over a certain size to have both green roofs and walls.[27]

Urban gardens

The benefits of urban farming and gardening are so apparent that the practice is spreading. The Green-Up program in The Bronx is a good example; this New York charity is turning vacant lots into community gardens, which are particularly popular with Puerto Rican and Jamaican immigrants, who often have rural origins.[28] Urban gardens are also very important resources for the unemployed, and New York City and Detroit are now providing thousands of acres of land so that the unemployed can grow their own food.[29] These initiatives are unlike the allotment system in the UK, because they are set up on a community basis rather than being given to individuals.

Community gardens often allow people to contribute in different ways, depending on their abilities and resources. The unemployed and retired can do much of the physical work, while busy working people can pay into the enterprise and pitch in with their labour when they have time. The harvest is shared between the investors of time and money, in a symbiotic system that benefits everybody. In addition, community gardens provide a place to meet, and a place where children can either be put to work or set free to play in a traffic-free environment. To top it off, gardening has been shown to lower the risk of dementia by 50%.[30]

Green energy

Another new development is the move towards decentralising energy supply. Nearly half of the money that goes into energy production actually goes to pay for transmission and infrastructure costs, which means that decentralising can save vast sums of money. In China, for example, research has suggested that if the government decentralised energy production they could save $400 million and cut the CO_2 output by 56%.[31] Such decentralisation projects have already taken place in the UK on a smaller, local basis. Woking Council, in Surrey, has created its own energy company with its own grid, with over 60 local generators which include gas-powered Combined Heat and Power (CHP) energy stations and large banks of solar panels on the roofs of municipal buildings.[32] This new, greener infrastructure supplies low carbon heating, cooling and electricity to all of the municipal buildings, plus more than 170 businesses and over 2000 households. Woking has cut its energy use by nearly half, and has cut its CO_2 emissions by 77% since 1990.[33]

The Netherlands are taking decentralisation even further, and 50% of their energy is now decentralised. The city of Rotterdam has added greenhouses to their power generation plants, which absorb CO_2 as well as supplying food for the city. San Francisco is also leading the way, with new legislation enacted in 2016 that requires ALL new buildings to be built with either solar panels or solar powered water heaters; other towns and cities in California are doing the same. Cheap, low tech solar water heaters were invented in Israel during an energy crisis in the 1950s, and since 1980 it has been a legal requirement for all new houses to have them installed. The notion is so logical and so cheap that they are now used throughout the Middle East (Figure 7.5).

Electric buses are another new technology that is cutting CO_2 emissions and creating cleaner cities. Electric buses are becoming more energy efficient, and there is even a new development called 'regenerative braking' that enables electric vehicles to create some of their own electricity every time the driver applies the brakes.[34] About 61 European cities are either testing or have adopted electric buses,[35] and a number of cities in the United States are also changing over to electric. New York City is looking into changing their fleet of 5700 buses for electric vehicles, at an estimated saving of $39,000 per year per bus, and a reduction of nearly 500,000 metric tons of CO_2 – even including the emissions that will be used to create the electricity.[36] If a greater reliance on renewable energy could be added to the mix, the carbon emissions will fall even further. In Los

Figure 7.5: Solar heaters (dudei Shemesh) in Jerusalem. They look like oil drums with a panel attached. Note also the typical domed architecture of the houses. (Photo © Jerry Bond)

Angeles, diesel buses have already been replaced by compressed natural gas (CNG), but the city's Transportation Authority plans to convert all 2300 buses to electric by 2030.[37]

The power to drive electric vehicles will be coming increasingly from renewable sources. The Netherlands are currently testing their first solar panel bicycle path, in which durable and slip-proof solar panels form a 76 m (250 foot) stretch of pathway which has generated enough electricity in six months to power a home for a year.[38] Solar panels for road surfaces are also in the development stage, and should soon be powering electric vehicles using solar car parks, driveways and roads.[39] Meanwhile, an Israeli start-up called ElectRoad is developing roads that will wirelessly recharge the batteries of electric cars and buses as they drive. This company can electrify a kilometre of road in just a single night, so we may be looking forward to a cleaner, quieter, greener world in the not-too-distant future.

Planning better cities

It has been estimated that the risk of depression is elevated by 40% just by living in a city, but we can reduce depression and improve our health by creating green cities with plenty of parks and trees.[40] We don't even have to give up huge swathes of land for parks: green roofs can become roof gardens, which can be oases of peace and greenery. Green roofs also, as we have seen, absorb pollutants and CO_2, emit oxygen, lower city temperatures in summer and also absorb noise. Green walls do all of this and more: in addition to all the green benefits of cooler summer temperatures and cleaner air, they provide an emotional uplift for passers-by on the street.

Green route ways between different areas of the city are another way of greening a city. Road rage incidents have been shown to decline where the travel route is greener, and people are three times more likely to walk along a green and landscaped pedestrian route than they are along unenhanced concrete.[41] Planting trees and shrubs along footpaths creates interest and makes urban walks a pleasure, and creating green corridors out of paths and bike paths along rivers and canals adds another dimension to the experience. Some towns are adding free food into the mix by planting community fruit trees and herb gardens in public spaces and along green route ways. We know that greening-up our cities and engaging with the natural world helps us to de-stress, but there is interesting new research that demonstrates this objectively, in controlled experiments. This work has led to the creation of a new term: biophilia.

Biophilia

The biologist E. O. Wilson noted the need for humans to 'affiliate' with other living organisms, and he has named this concept 'biophilia', which he describes as 'the innate tendency [of humans] to focus on life and lifelike processes'.[42] 'The more we know of other forms of life,' he wrote, 'the more we enjoy and respect ourselves.' Wilson published his biophilia hypothesis in 1984, at a time when the notion that humans have a deep connection to the natural world was making a comeback. It is bizarre to think that we ever 'forgot' this connection, but the disconnect came about in the late 19th and early 20th century when we didn't have the scientific methods available to prove the mind-nature-body connection.

Doctors in the 19th and early 20th century had recognised that the natural world offers relief from the noise, pollution and overcrowding that had increasingly become the norm following the industrial revolution, and doctors prescribed exposure to nature as a cure for frayed nerves. Sanatoriums were a big industry, and were located in beautiful places, often in woodlands. In the 19th century, doctors in Germany and America noticed that sanatoriums for tubercular patients were more effective when placed in forests, as opposed to open landscapes, but medical science at the time was unable to provide the link to explain it, and sanatoriums fell out of fashion.

All that changed in the 1980s, when researchers began to find objective ways to test the mind-body link. The renewed interest was spearheaded by Roger Ulrich, a geography PhD student in Michigan. Ulrich noticed that local people were taking a long and scenic drive rather than a short journey on the motorway (freeway) in order to get to a local mall, and he wanted to find out why they were prepared to sacrifice time and petrol for something that our modern economic theory does not regard as 'valuable'. In 1979 he set up his first test, for which he used 46 stressed-out students who had just completed an exam. He divided the group in two, and showed one group a series of slides depicting the natural world, while the other group saw slides of clean and orderly urban scenes. Those who saw the natural scenes were relaxed, happy, playful and affectionate, while those who watched the slides of urban scenes responded with sadness and depression.[43] So far, so predictable – but then Ulrich began a series of objective tests. He set up an electro-encephalograph (EEG) to measure brain activity in the two groups, again looking at urban or natural scenes. The group exposed to scenes of nature showed higher alpha waves, which are associated with higher levels of serotonin, which is linked with happiness. The group that saw urban scenes had lower alpha waves and more pronounced beta waves, which are linked with anxiety. The tests that followed included testing the pulse of the subjects using an electrocardiogram, then a test for stress which analysed skin conductance (which measures tiny amounts of sweat on the skin), and then a test for muscle tension using electromyography (EMG). All the results confirmed his initial findings: viewing scenes from nature acts as 'visual valium'.[44]

Ulrich's studies were replicated in studies around the world, showing that human reactions to nature seem to be consistent across cultures. Other tests followed: the hormone cortisol is linked with stress, and therefore cortisol levels can be measured as an objective marker for emotional experience. Haemoglobin levels are another independent indicator for stress. Magnetic resonance imaging showed that viewing natural scenes affects the dopamine reward centre in the brain, which is another objective way to show that nature makes us happy.

These studies are only the beginning. Pictures of nature lower stress, but actually being in nature does the same, with a number of other added benefits. In 1982 the Forest Agency in Japan wished to encourage walkers into the woods, and so they created a number of walking trails. They also coined the term 'Shinrin Yoku', 'shinrin' meaning forest, and 'yoku' meaning to bathe or to bask, to attract people into the woods. The term is generally translated as 'forest bathing', and studies have shown that it is, in fact, remarkably effective as a treatment for stress, and it even improves our resistance to disease.[45] A simple walk in the woods can significantly reduce cortisol levels and other measures of stress, while also lowering levels of depression and anger. It also improves sleep, lowers blood pressure, and has an unexpected beneficial effect on the immune system: forest walks actually help us fight disease by increasing the number of 'natural killer cells' (NK cells) in our bloodstream. NK cells are white blood cells that fight both viruses and tumours, and it has been demonstrated that the beneficial effect we gain from the forest comes in part from breathing in the volatile organic substances, called 'phytoncides' (or essential oils expressed by trees) which are the trees' natural antimicrobial defence.[46] Further studies have shown that while exercise is in itself beneficial, exercising in a natural environment is significantly more beneficial than exercising in an urban setting.[47]

The evidence is stacking up to show that humans thrive in a natural environment and are likely to feel more depressed in an urban setting – but in 2014, 54% of the world's population lived in cities,[48] while in 100 years that figure is expected to rise to 60–70%.[49] There are two approaches to this problem, and I recommend both: 1) we need to make our countryside more liveable and more connected, in order to stop the drift to cities, and 2) we need to make our cities greener, healthier, more liveable, and also more productive and more self-sufficient.

Symbiotic mutualism

I keep returning to the notion of symbiosis – a biological concept that means 'living together'. Symbiotic relationships are 'mutualistic' when both organisms benefit from the other, and an example of symbiotic mutualism is the relationship between plants and fungi: more than 48% of land plants rely on fungi living in their root systems. The plants provide a host for the fungi, and the fungi take up nutrients and convert them into a form that can be taken up by the host plants. The relationship between bees and flowers is also mutualistic, as is the notion of the 'Three Sisters' (maize, beans and squash) which I discussed in Chapter 4.

Symbiosis is engagingly discussed in the award-winning 1974 bestseller, *The Lives of a Cell*. This was written by a doctor and medical researcher, Lewis Thomas, and in this remarkable collection of essays he argued that most interactions in the natural world are symbiotically mutualistic – that is, they are interactions in which both organisms benefit. I think Thomas hit on something important in his essays, and at the time his notion was a huge departure from the more common perception of all living things being in a state of constant warfare. It is interesting to view the world in this way, because when we look for the ways in which one organism assists another, we begin to look for ways to encourage and increase this mutual aid. We view the world through the lens of our own culture, but anthropology teaches us that there are many different ways of seeing and understanding what is around us. One of the things Thomas did, in his own small way, was to change the way our own culture views the world. This new thinking has helped us to think about more cooperative ways forward.

Today, many politicians – particularly in America – are presenting our options as either/or decisions. If we protect the environment, they tell us, it will lead to starvation and will destroy the economy. We are told that the only way to foster continued prosperity is to continue with an economy based on fossil fuel. I and many others would argue that investment in a new green economy is another way to achieve prosperity, but with this difference: by preserving the natural resources on which we depend, our prosperity can be sustainable, so that future generations can also enjoy a comfortable life – and they can enjoy it in a greener and richer environment. Where governments invest in new, green technology, and where legislation encourages the development of such technology, we are seeing huge technological leaps forward, a vast range of new investment opportunities and the happy outcome of a greener world. My point is that there don't always *have* to be winners and losers. We have the technological and economic capacity to ensure that nearly everybody wins, if only we could change our outlook and think about the problems in a mutualistic way. We don't have to make huge sacrifices in order to have a greener world; we don't have to decide 'which 50 million people will starve'; we just need a bit of imagination.

When we look at the symbiotic mutualism that takes place in the natural world around us, it is possible for us to start seeing the world as a less threatening place, and not purely as 'nature red in tooth and claw'. I'm not saying we should ignore the savagery of predator-prey relationships, which are also part of the natural world, and I don't want to diminish the fact that there are horrible diseases and venomous snakes and insects that want to lay eggs in your brain, but if we focus our attention on all the mutually beneficial relationships in nature, we can look for ways in which we, too, can join in with this system of mutual aid. Mutualism is a theme that runs throughout this book, and it is a useful concept to think about if we wish to create a greener, more sustainable and more equitable world.

Social continuity and change

A fundamental principle of Capitalist Economics is that people have unlimited wants. When I was an undergraduate we were taught that this is a universal truth, and yet it disregards everything we know about anthropology, and it disregards the fact that in other cultures people have different values. Hunter-gatherers, for instance, are traditionally nomadic people who have very few possessions, and who would simply be encumbered if they were to accumulate them. In Buddhist and Taoist belief, it is thought that excess possessions make people unhappy. Taking this a step further, in Buddhist economic theory it is assumed that you need just enough to live comfortably, because poverty is demeaning and unhealthy, and furthermore, people need enough extra so that they can give to others, because that (on the whole) creates a sense of well-being for the giver, and it also reinforces social bonds.

While we would all like to believe that this ideal of human behaviour could come about effortlessly, we know that in practice we are a mixed bunch. Some of us are prepared to make huge sacrifices for the greater good, but most of us are unwilling to go to extremes of sacrifice and asceticism. (When I was a little girl, my father used to exhort me to 'be a Spartan', whenever I was being forced to endure some physical discomfort. I'm told I once lost patience and snarled, 'I'd rather be a Sybarite'.) Most of us function on a fairly materialistic and Sybaritic (*i.e.* self-indulgent) level, which means that we are inclined to make sacrifices for the Greater Good only if there is something in it for us. This doesn't necessarily mean that we are all in a competition to gobble up the world's resources; rather, it means that we need to agree government regulations to incentivise businesses, technologies and practices that are beneficial for all of us and good for the planet. This may sound like wishful thinking, but large and beneficial changes can be brought about by small prompts to change our behaviour. To do this well, we have to understand what really motivates people.

The new field of behavioural science shows that people do not always behave rationally – hardly a surprise, but since economists base their models on the notion that people are rational actors, it is important to know when and how we fail to act predictably. The fact is, we often put off doing things that would benefit us, not because we are lazy, exactly, but because we feel overwhelmed by the number of obstacles that stand in our way. In other words, we become daunted by the size of the task ahead, and put it off until later. Here is an example. The UK government offered householders money to encourage us to put in loft insulation, which – if we actually got around to doing it – would save us energy and lower our bills. What was stopping people from taking up this fantastic opportunity? It turns out that most of us acquire large amounts of clutter in our attics or lofts, and decluttering is a daunting task because it's a big job and it involves a huge amount of decision-making, which is stressful and exhausting. The government's 'Behavioural Insights Team', otherwise known as the 'Nudge Unit', found that there was a significant increase in the take-up of loft insulation when they offered a combined loft-clearing and insulation service.[50] It cost householders more money, but reduced the hassle and saved people time – and it seems we really value our time.

The 'Nudge Unit' has been derided and sometimes regarded with suspicion, but what's not to like about finding ways to save money *and* help the environment? This group actually managed to increase recycling by 30%, just by telling people in leaflets that 'most people recycle responsibly'.[51] It turns out that most of us really do want to be good citizens, or at least we want to be as good as our neighbours – and dropping the hint that 'most people' do this or that is a more powerful 'nudge' than fining people.

Top-down, bottom-up or both?

Making decisions about how we want to govern our own countries is one thing, but how do we influence others? The notion that a charity, or 'non-government organisation' (NGO) should go to work in a developing country and take charge of the 'natives' is no longer acceptable.[52] Not only is it condescending, but it is often the case that the people involved do not fully understand the culture they are trying to assist, and therefore make serious mistakes – so inappropriate, sometimes, as to negate their whole project. We saw an example of this when aid workers built houses in the form of tombs for refugees in Africa, and the refugees preferred to live outside rather than dwell in what they regarded as mausoleums. Another example was when aid workers sank wells in the desert, but made the pump action too heavy for the women to manage. If they had simply spoken to the people about their needs, it would have become evident that collecting water was regarded as 'women's work', and therefore the pumps must be operable by women. The issue of whether or not the culture ought to change may or may not be something else to consider, but if it is addressed, it needs to be done with two-way conversation and sensitivity; people don't often thank you for coming into their country and telling them how to live their lives (although sometimes they do – which is why it has to be a conversation[53]).

It is essential that NGOs talk to the people they are trying to assist, but bottom-up changes can also be very effective. Grassroots organisations bring local knowledge to local problems, although they don't always bring the necessary organisational skills or fundraising expertise. There have been many very successful efforts, however, and as people become more skilled at networking, grassroots organisations are working transformative wonders around the world. A story that particularly impressed me was one that took place in Husa, Sweden. This remote village nearly died a death when the copper mine shut down, and by 1979 there were just 90 people left.[54] The village shop had closed, and the school looked likely to shut down too, as there were few remaining young families. Then, a local artist organised a group of people who got together and wrote a play about the village – and things took off from there. People were inspired, and they started forming co-ops. At least fifteen village associations have been formed since the 1970s, and co-ops formed to build houses for locals and holiday homes to let to vacationers. One co-op started up a saw-mill, and another created a museum and restaurant, while another was organised to keep the roads in good repair. Many other new businesses formed, including a goat farm, a bakery, a ski resort and a pony-trekking business. By the mid-1990s the population was up to 160, including 40 children, which just goes to show that young people DO want to live in the country; the city has its attractions, but so does raising children in a dynamic and vibrant village.

The 'bottom-up' approach is well known in Latin America, where the Farmer to Farmer program (Campesino a Campesino), was formed by hundreds of thousands of small farmers in the 1990s. The basis of this organisation is farmers helping themselves and each other by sharing knowledge of sustainable farming methods. The program has led to farmers tripling and even quadrupling their yields, and the use of cover crops reduced the time needed for weeding by 75%.[55] Ecologically and economically, the movement was a huge success.

Unfortunately, the Campesino a Campesino group is struggling against government systems that do not support their production methods, and there has now been a call for assistance from NGOs, who can help to advocate for the farmers and who can teach the farmers to fight for themselves more effectively. This is quite possibly the best way forward: a partnership between local people and NGOs, or alternatively a partnership directly between grassroots organisations and governments. Governments or city councils can provide farmers with credit, seeds and information, and they can also provide services. In Cuba, the government stepped in and offered small plots of land to independent farmers, and they also provide compost to farmers in the cities. This is a particularly useful service, because the city farmers have limited space, but there is a city full of waste that can be drawn upon. This is an idea that is catching on more widely. New York City's sanitation department has instituted a curb-side food waste collection program, which at the moment collects organic waste from 3.3 million residents.[56] The waste is used to produce compost, which is given out to individual citizens and also to the many community farms and gardens within the city. The waste is also used to produce biogas, which – in the near future – will provide vehicle fuel and electricity for city residents.

Europeans may wonder about America's commitment to sustainability, but individual cities are making independent progress towards sustainable development by introducing electric buses, green roofs, city gardens and urban composting schemes. Don't write off America because of the national policies, which may change with each presidency. Look instead at what municipalities like New York and San Francisco are doing to promote sustainable development and to fight climate change.

Peak oil

In the 1950s a geophysicist for Shell Oil, M. K. Hubbert, predicted that, since oil reserves are finite, the year of peak oil production in America would occur in 1965, or (should we find more major oil sources) in 1970.[57] His study was based on his discovery that oil production follows a bell-shaped curve, in which recovery of oil increases rapidly on the discovery of a new oil field, and then gradually declines. No one took his analysis seriously until 1970, when oil production in America did in fact begin to decline. Hubbert's methods were later used to predict global peak oil, and that calculation was also about five years off the mark, as it was suggested that we would reach peak oil in the year 2000, and the actual year of peak oil production was (arguably) 2005.[58] Hubbert's detractors believe that five years is a major margin of error which indicates a fatal flaw in his science, and it has also been noted that the bell-shaped curve of oil production is not smooth, but has many minor peaks and troughs – but these seem like niggling arguments when you consider that the man's predictions for this major world event were just five years out. Although new oil reserves are still occasionally discovered – and we are developing new technologies to gain access to the more inaccessible sources – it is now widely believed that we have already reached peak oil.

What does this mean for the future? Oil reserves are probably now in decline, and it will take tens of millions of years for new oil to form (or some say hundreds of millions). We are going to have to learn to live without oil. However, we are not going to have to switch to an oil-free economy overnight, as Cuba practically had to do. We have time to develop alternatives, and we have the technology. We also have good reason to do it as soon as possible: CO_2 in the atmosphere is at dangerous levels, and the effects are all around us.

Biodiversity and climate change

The World Wildlife Fund estimates that between 1970 and 2012, worldwide animal numbers dropped by 58%.[59] By the year 2020, they reckon that we will have lost 67%. At the moment, the oceans are bearing the brunt of this ecological decline, because 90% of the increased heat is being absorbed by the world's oceans – which in places have warmed by 1°C. That isn't all that's happening. The oceans are also becoming more acidic, which is bad news for most marine life apart from jellyfish.[60]

Biologists have now estimated that 20% of the world's coral reefs are dead, but 93% are showing signs of bleaching, which is the first sign of decline.[61] Corals live symbiotically with algae, which provide energy and nutrients to the coral, as well as giving them their glorious colours. The increased warmth of the oceans is causing corals to expel the algae, hence the bleaching and subsequent death of our once beautiful reefs. Following a recent survey, one despondent biologist said, 'this was the saddest research trip of my life'.[62]

The problem extends beyond the loss of life, colour and biodiversity in our oceans; it is an economic problem too, because around 500 million people depend on coral reefs for their livelihoods – reefs provide fish, coastal protection and also a substantial sum of money from tourism. Lose the reefs, and you lose an ecosystem that 30 million Pacific Islanders depend upon. However, their problems don't end there: at least eight Pacific islands have been submerged by rising sea level, and the high salt water table is also salinating the water supply on many islands. High tides are making life dangerous and uncertain on a number of islands, but at the moment the United Nations does not give refugee status to people whose homes are no longer habitable due to climate change – so where will these people go? Sea levels are predicted to rise by about 1 m by 2100, using moderately conservative estimates,[63] and probably by 2 m per century in the following years – which means that not only Pacific islands will be submerged, but also much of Bangladesh and other low-lying coastal countries, as well as cities such as Miami and New Orleans. If the Greenland and West Antarctic ice sheets melt, then global sea level will rise by about 10 m.[64]

The Fifth Assessment Report of the UN Intergovernmental Panel on Climate Change (IPCC) stated that:

> From 1880 to 2012, the average global temperature increased by 0.85°C. Oceans have warmed, the amounts of snow and ice have diminished and the sea level has risen. From 1901 to 2010, the global average sea level rose by 19 cm as oceans expanded due to warming and ice melted. The sea ice extent in the Arctic has shrunk in every successive decade since 1979...
>
> Given current concentrations and ongoing emissions of greenhouse gases, it is likely that the end of this century will see a 1–2°C increase in global mean temperature above the 1990 level... The world's oceans will warm and ice melt will continue. Average sea level rise is predicted to be 24–30 cm by 2065 and 40–63 cm by 2100... Most aspects of climate change will persist for many centuries, even if emissions are stopped.[65]

Here is another worry. Much of the planet – around 20% – is covered in permafrost. This is now melting. Not only is this causing boreal forests to collapse, but it is also releasing some of the millions of tons of CO_2 that was, until recently, stored in this frozen land.[66] The problem is this: if we stop all carbon and methane emissions today, the knock-on effects will carry on for years to come – and if the permafrost melts, we will see the effects catapulting out of our control. We need to act *now*.

The future

I have mostly taken an optimistic tone throughout this book, firstly because I am inspired by the sustainable technologies of the past and present, and secondly because I believe there is nothing to be gained by throwing up our hands in despair and declaring that the problem is too big for us to solve. There is another thing that gives me hope and optimism, and it comes from my years as a university lecturer. Many years ago, I observed to a colleague that we can learn things from the youngest undergraduates, as well as the most senior professors. Sometimes one of my students will make a really interesting observation, or will see something in a different light, or my students find new ways of expressing ideas. What is most remarkable, though, is that young people do not know what is and isn't possible. This leads them to attempt things that we weary older folk know to be unfeasible – and the blindingly wonderful thing is that they sometimes achieve these impossible goals. When you combine optimism, enthusiasm and a determination to succeed, astonishing things can be accomplished.

There are so many success stories in the recent past. If you are old enough, reflect back to the 1970s, when people who cared about the environment were regarded as cranks and eccentrics. Perhaps we are still regarded as cranks and eccentrics, but we have achieved a great deal. Rivers in most of the Western world are no longer so polluted that they catch fire, and here in the UK we are seeing many native species coming back from the brink of local extinction. The fight to protect our environment never ends, because new technologies cause new kinds of pollution and environmental damage, but there are reasons to be optimistic. Here in Wales, there are various environmental protection schemes in action, and just walking through the fields I notice that many wildflower species are making a comeback, and there is a rise in the number of grass species in the meadows. Red kites – once very near to extinction – fly up and down our valley, while dippers, grey wagtails, kingfishers and the occasional heron hunt in the clear streams that run off the mountain behind our house.

Figure 7.6: Pembrokeshire heather and gorse. (Photo © Jerry Bond)

Figure 7.7: The author at St Brides, Pembrokeshire – 'the road goes ever on and on'. (Photo © Jerry Bond)

I live my life in the countryside of west Wales, where I am surrounded by rolling green hills and hedgerows filled with bluebells and red campion, as well as wild strawberries, blackberries and bilberries that provide a late summer feast. I go walking with my husband every few weeks on the Pembrokeshire Coast Path – voted by Lonely Planet guides as one of the most beautiful places in the world – and on a spring day, we can count around 80 different wildflower species all blooming at once, completely carpeting the cliffs overlooking the sea (Figures 7.6 and 7.7). We stare at the seals, who stare right back at us, and we watch the busy guillemots nesting in the spring. The blue butterflies are a joy, and we sit and watch the gannets diving, and listen to the skylarks singing their little hearts out in the spring. I love this land with all my heart, and I know it with the time-depth of a practicing archaeologist. When we're out walking we notice the remains of the Iron Age hillforts and promontory forts on the coast, and the signs of long-gone hedgerows. We frequently spot the flint tools of our prehistoric ancestors.

My readers will have landscapes that you, too, know and love with all your heart. Wherever you are in the world, you probably treasure your countryside as much as I treasure mine. Isn't it worth fighting for? Climate change threatens all of us, which means that we have a whole world to fight for. Saving it won't be achieved by ignoring the problems or by giving up because it's too difficult. We are indeed facing an overwhelming task, but we need to face it together, fearlessly, and with optimism.

The time to despair is *never*.

Notes

1 http://www.kkl-jnf.org/people-and-environment/kkl-jnf-projects-partners/dfu-2013/carmel-ancient-agriculture-terrace/ Accessed 9 November 2018.
2 https://www.smithsonianmag.com/history/farming-like-the-incas-70263217/ Accessed 9 November 2018.
3 https://qcat.wocat.net/af/wocat/approaches/view/approaches_2635/ Accessed 9 November 2018.
4 Tarolli 2014; http://palatepress.com/2016/10/wine/terraced-vineyards-in-jeopardy/ Accessed 9 November 2018.
5 https://www.israel21c.org/could-neotops-water-spheres-save-california/?utm_content=buffer41fdd&utm_medium=social&utm_source=facebook.com&utm_campaign=buffer Accessed 9 November 2018.
6 https://www.israel21c.org/solar-powered-desalination-cuts-energy-costs-by-90/ Accessed 9 November 2018.
7 Awawdeh *et al.* 2012.
8 Abdel-Shafy *et al.* 2010.
9 Albalawneh *et al.* 2015.
10 Pretty 1998a, 100.
11 Pretty 1998a, 103.
12 Pretty 1998a, 85.
13 Harlan 2008.
14 Barker and Coutts 2016.
15 Barker and Coutts 2016, 96.
16 Scholz 2013.
17 Scholz 2013.
18 Garrison and Horowitz 2012.
19 Bathurst *et al.* 2010.
20 Bathurst *et al.* 2010; Jim 2017.
21 Cantor 2008, 33.
22 Cardiff News (December 2007). Keep cool with a green roof. Page 5.
23 Cantor 2008, 33; Garrison and Horowitz 2012.
24 Cantor 2008, 20; Barozzi *et al.* 2016.
25 Cantor 2008, 33.
26 Cantor 2008, 88.
27 Cantor 2008, 19.
28 Girardet 1999, 55.
29 Girardet 1999, 55.
30 Selhub and Logan 2012.
31 http://www.no2nuclearpower.org.uk/reports/Decentralised_Energy.pdf Accessed 9 November 2018.
32 Greenpeace 2005, 18.
33 Greenpeace 2005, 18.
34 https://www.globalmasstransit.net/archive.php?id=17689 Accessed 9 November 2018.
35 https://www.theagilityeffect.com/en/article/electric-buses-set-conquer-world/ Accessed 9 November 2018.
36 Aber 2016.
37 http://www.thedrive.com/tech/17073/los-angeles-aims-for-all-electric-bus-fleet-by-2030 Accessed 9 November 2018.
38 https://www.triplepundit.com/2016/01/looking-at-the-dutch-solar-bike-path-after-one-year/ Accessed 9 November 2018.

39 http://www.solarroadways.com/ Accessed 9 November 2018.
40 Byrne 2017.
41 Farr 2008, 49.
42 Wilson 1984; Kellert and Wilson 1993.
43 Selhub and Logan 2012, 14.
44 Selhub and Logan 2012, 16.
45 Park *et al.* 2010; Qing 2010.
46 Qing 2010.
47 Qing 2010.
48 World Health Organisation Global Health Observatory (GHO) data, available online at: http://www.who.int/gho/urban_health/situation_trends/urban_population_growth_text/en/ Accessed 9 November 2018.
49 Girardet 1999.
50 Halpern 2015, 4.
51 Halpern 2015, 35.
52 For a discussion of 'Equitable Resilience', see Matin *et al.* 2018. This article discusses the notion that not only do you have to discuss people's needs directly with the people involved, but you also have to take account of all strata within the society you aim to assist. The government or leaders of a country will not necessarily be interested in supporting all sectors of society, and the article discusses the need to listen to the position of those who have 'differential access to power, knowledge, and resources'.
53 Helping women to become more economically independent, for instance, has a large and beneficial impact on a country's productivity, prosperity and child health. Increasing women's participation in the economies in developing countries is therefore one of the United Nations' goals in the new 2030 Agenda for Sustainable Development (FAO 2017a).
54 Pretty 1998a, 227.
55 Altieri *et al.* 2012.
56 http://www1.nyc.gov/assets/dsny/site/services/food-scraps-and-yard-waste-page/overview-residents-organics Accessed 9 November 2018.
57 Deffeyes 2001, 1.
58 Deffeyes 2001, 1.
59 WWF 2016, 12.
60 Maslen 2017, 44.
61 Maslen 2017, 47.
62 Maslen 2017, 47.
63 Mann and Kump 2015, 158.
64 Mann and Kump 2015, 158.
65 http://www.un.org/en/sections/issues-depth/climate-change/ Accessed 9 November 2018.
66 IPCC 2014.

Bibliography

Abdalla, M., Osborne, B., Lanigan, G., Forristal, D., Williams, M., Smith, P. and Jones, M. B. (2013). Conservation tillage systems: a review of its consequences for greenhouse gas emissions. *Soil Use and Management* 29, 199–209.

Abdel-Shafy, H. I., El-Saharty, A. A., Regelsberger, M. and Platzer, C. (2010). Rainwater in Egypt: quantity, distribution and harvesting. *Mediterranean Marine Science* 11(2), 245–257.

Aber, J. (2016). Electric Bus Analysis for New York City Transit. Available online at: http://www.columbia.edu/~ja3041/Electric%20Bus%20Analysis%20for%20NYC%20Transit%20by%20J%20Aber%20Columbia%20University%20-%20May%202016.pdf (accessed 13 November 2018).

Adeel, Z., Schuster, B. and Bigas, H. (eds) (2008). *What Makes Traditional Technologies Tick? A Review of Traditional Approaches for Water Management in Drylands.* Hamilton, Ontario: The United Nations University.

Adhikary, N. (2016). Vernacular architecture in post-earthquake Nepal. *International Journal of Environmental Studies* 73(4), 533–540.

Afrol News (no date). *Water harvesting promoted in Troubled Karamoja Cluster.* Available online at: http://www.afrol.com/articles/12946 (accessed 13 November 2005).

Agarwal, A. (2001). The value of natural capital. *Development Outreach*, Winter 2001.

Agarwal, A. and Narain, S. (1992). Traditional systems of water-harvesting and agroforestry. In G. Sen (ed.), *Indigenous Vision. Peoples of India: Attitudes to the Environment.* New Delhi: India International Centre, Sage Publications.

Ahmad, Z. (2017). Why modern mortar crumbles, but Roman concrete lasts millennia. *Science* 358(6359), 6 October 2017.

Al-Bakri, J. T., Ajlouni, M. and Abu-Zanat, M. (2008). Incorporating land use mapping and participation in Jordan. *Mountain Research and Development* 28(1), 49–57.

Albalawneh, A., Chang, T.-K., Huang, C.-W. and Mazareh, S. (2015). Using landscape metrics analysis and analytic hierarchy process to assess water harvesting potential sites in Jordan. *Environments* 2(3), 415–434.

Al-Ghafri, A., Inoue, T. and Nagasawa, T. (2003). Daudi Aflaj: the qanats of Oman. The Third Symposium on Xinjang Uyghur, Chipa, Japan. Available online at: https://www.researchgate.net/publication/259931405_Daudi_Aflaj_the_Qanats_of_Oman (accessed 13 November 2018).

Al-Hebshi, A. (2005). The role of terraces management on land and water conservation in Yemen (A case study in Kuhlan-Affar/Wide Sharis Districts). Paper presented at the Third International Conference on Wadi Hydrology, Sana'a, Yemen, 12–14 December 2005.

Allen, M. S. (2004). Bet-hedging strategies, agricultural change, and unpredictable environments: historical development of dryland agriculture in Kona, Hawaii. *Journal of Anthropological Archaeology* 23(2), 196–224.

Allen, T. G. and Robinson, M. A. (1993). *The Prehistoric and Iron Age Enclosed Settlement at Mingies Ditch, Hardwick-with-Yelford, Oxford.* Thames Valley Landscapes: The Windrush Valley, Vol. 2. Oxford: Oxbow Books.

Altieri, M. A. (2002). Agroecology: the science of natural resource management for poor farmers in marginal environments. *Agriculture, Ecosystems and Environment* 1971, 1–24.

Altieri, M. A., Anderson, M. K. and Merrick, L. C. (1987). Peasant agriculture and the conservation of crop and wild plant resources. *Conservation Biology* 1(1), 49–58.

Altieri, M. A., Funes-Monzote, F. R. and Petersen, P. (2012). Agroecologically efficient agricultural systems for smallholder farmers: contributions to food sovereignty. *Agronomy for Sustainable Development* 32(1), 1–13.

Avni, G., Porat, N. and Avni, Y. (2013). Byzantine-Early Islamic agricultural systems in the Negev Highlands: stages of development as interpreted through OSL dating. *Journal of Field Archaeology* 38(4), 332–346.

Awawdeh, M., Al-Shraideh, S., Al-Qudah, K. and Jaradat, R. (2012). Rainwater harvesting assessment for a small size urban area in Jordan. *International Journal of Water Resources and Environmental Engineering* 4(12), 415–422.

Aw-Hassan, A., Bruggeman, A. and Ebrahim, A. R. Y. (2002). The Yemeni mountain terraces project: institutionalising community-based integrated natural resources management research. *Caravan* 16, 24–26.

Bainbridge, D. A. (2001). Buried clay pot irrigation: a little known but very efficient traditional method of irrigation. *Agricultural Water Management* 48, 79–88.

Baker, C. J. and Saxton, K. E. (2007). The 'what' and the 'why' of no-tillage farming. In C. J. Baker, K. E. Saxton, W. R. Ritchie, W. C. T. Chamen, D. C. Reicosky, F. Ribiero, S. E. Justice and P. R. Hobbs (eds), *No-Tillage Seeding in Conservation Agriculture* (2nd edition), 1–10. Wallingford: CAB International and FAO. Available online at: http://www.fao.org/docrep/012/al298e/al298e00.htm (accessed 14 November 2018).

Barclay, G. J. (1997). The Neolithic. In K. J. Edwards and I. B. M. Ralston (eds), *Scotland: Environment and Archaeology, 8000 BC–AD 1000*, 127–49. Chichester: Wiley.

Barker, G. (1996). *Farming the Desert: The UNESCO Libyan Valleys Archaeological Survey.* London: UNESCO.

Barker, G. (2000). Farmers, herders and miners in the Wadi Faynan, southern Jordan: a 10,000 year landscape archaeology. In G. Barker and D. Gilbertson (eds), *The Archaeology of Drylands: Living at the margin*, 63–85. London and New York: Routledge.

Barker, R. and Coutts, R. (2016). *Aquatecture: buildings and cities designed to live and work with water.* Newcastle: RIBA Publishing.

Barozzi, B., Bellazzi, A. and Pollastro, M. C. (2016). The energy impact in buildings of vegetative solutions for extensive green roofs in temperate climates. *Buildings* 6(3), 33. Available online at: https://www.mdpi.com/2075-5309/6/3/33 (accessed 14 November 2018).

Bathurst, R. R., Zori, D. and Byock, J. (2010). B Diatoms as bioindicators of site use: locating turf structures from the Viking Age. *Journal of Archaeological Science* 37, 2920–2928

Bean, R., Olesen, B. W. and Kim, K. W. (2010). History of Radiant Heating & Cooling Systems, Part 1. *American Society of Heating, Refrigerating and Air-Conditioning Engineers (ASHRAE) Journal* 2010, 40–47.

Beg, M. S. (2016). Issues of conservation and adaptation in protecting Kashmir's vernacular heritage. *International Journal of Environmental Studies* 73(4), 524–532.

Behre, K.-E. (1992). The history of rye cultivation in Europe. *Vegetation History and Archaeobotany* 1, 141–156.

Bell, M. and Boardman, J. (1992). *Past and Present Soil Erosion.* Oxbow Monograph 22. Oxford: Oxbow Books.

Benachour, N. and Séralini, G. E. (2009). Glyphosate formulations induce apoptosis and necrosis in human umbilical, embryonic, and placental cells. *Chemical Research in Toxicology* 22(1), 97–105.

Benbrook, C. (2009). *Impacts of Genetically Engineered Crops on Pesticide Use: The first thirteen years.* Critical Issue Report: The First Thirteen Years. Available online at: https://link.springer.com/article/10.1186/2190-4715-24-24 (accessed 14 November 2018).

Bennett, W. H. (1911). *The Moabite Stone.* Edinburgh: T & T Clark.

Best, E. P. H., Verhoeven, J. T. A. and Wolff, W. J. (1993). The ecology of The Netherlands wetlands: characteristics, threats, prospects and perspectives for ecological research. *Hydrobiologia* 265, 305–320.

Boersma, J. (2005). Dwelling mounds on the salt marshes: the terpen of Friesland and Groeningen. In L. P. Louwe Kooijmans, R. W. van den Broeke, H. Fokkens and A. L. van Gijn (eds), *The Prehistory of the Netherlands* Vol. 2, 557–560. Amsterdam: Amsterdam University Press.

Boixadera, J., Poch, R. M., García-González, M. T. and Vizcayno, C. (2003). Hydromorphic and clay-related processes in soils from the Llanos de Moxos (northern Bolivia). *Catena* 54, 403–424.

Bradley, R. (1978). Prehistoric field systems in Britain and northwest Europe: a review of some recent work. *World Archaeology* 9, 265–280.

Brady, N. C. and Weil, R. R. (1999). *The Nature and Properties of Soils.* Upper Saddle River, N.J.: Prentice-Hall, Inc.

Briones, A. M. (2012). The secrets of El Dorado viewed through a microbial perspective. *Frontiers in Microbiology* 3, 1–6.

Brouwer, R., van Ek, R., Boeters, R. and Bouma, J. (2001). *Living with Floods: an integrated assessment of land use changes and floodplain restoration as alternative flood protection measures in the Netherlands.* Centre for Social and Economic Research on the Global Environment, Working Paper ECM 01-06. Norwich: University of East Anglia.

Brown, A. G. (1997). *Alluvial Geoarchaeology: floodplain archaeology and environmental change.* Cambridge: Cambridge University Press.

Brown, B. J. and Marten, G. G. (1986). The ecology of traditional pest management in Southeast Asia. In G. G. Marten (ed.), *Traditional Agriculture in Southeast Asia: a human ecology,* 241–272. Boulder, CO: Westview Press.

Bruins, H. J. and Van Der Plicht, J. (2005). Desert Settlement through the Iron Age: Radiocarbon dates from Sinai and the Negev Highlands. In T. E. Levy and T. Higham (eds), *The Bible and Radiocarbon Dating,* 349–366. London: Equinox.

Brush, S. B. (1980). The environment and native Andean agriculture. *America Indigena* 40, 161–172.

Byatnal, A. (2012). Study questions sustainability of Bt cotton in water-starved Vidarbha. *The Hindu,* June 24, 2012. Available online at: https://www.thehindu.com/news/national/study-questions-sustainability-of-bt-cotton-in-waterstarved-vidarbha/article3563411.ece (accessed 14 November 2018).

Byrne, J. (2017). Planners know depressingly little about a city's impacts on our mental health. The Conversation, July 31, 2017. Available online at: https://theconversation.com/planners-know-depressingly-little-about-a-citys-impacts-on-our-mental-health-81098 (accessed 14 November 2018).

Cantor, S. L. (2008). *Green Roofs in Sustainable Design.* New York and London: W. W. Norton & Co., Inc.

Capel, P. and Capelli, K. (2011). Widely used herbicide commonly found in rain and streams in the Mississippi river basin. United States Geological Survey Technical Announcement, released 29 August 2011. Available online at: https://archive.usgs.gov/archive/sites/www.usgs.gov/newsroom/article.asp-ID=2909.html (accessed 14 November 2018).

Capistrano, A. D. and Marten, G. G. (1986). Agriculture in Southeast Asia. In G. G. Martin (ed.), *Traditional Agriculture in Southeast Asia: A human ecology perspective,* 6–19. Boulder, CO: Westview Press.

Carson, R. (1962). *Silent Spring.* Boston and New York: Houghton Mifflin.

Chang, F. C., Simcik, M. F. and Capel, P. D. (2011). Occurrence and fate of the herbicide glyphosate and its degradate aminomethylphosphonic acid in the atmosphere. *Environmental Toxicology and Chemistry* 30(3), 548–555.

Ciriacono, S. (2006). *Building on Water: Venice, Holland and the construction of the European Landscape in early Modern Times.* New York and Oxford: Berghahn Books.

Cleveland, D. A., Soleri, D. and Smith, S. E., (1994). Do folk crop varieties have a role in sustainable agriculture? *BioScience* 44(11), 740–751.

Coe, M. D. (1964). The chinampas of Mexico. *Scientific American* 211, 90–98.

Coghlan, A. (2006). China's GM cotton battles a new bug. *New Scientist* 29 July, 2006, 5.

Coles, B. and Coles, J. (1989). *People of the Wetlands: bogs, bodies and lake-dwellers*. London: Thames and Hudson.

Conway, G. (2005). The doubly green revolution. In J. Pretty (ed.), *The Earthscan Reader in Sustainable Agriculture*, 115–127. London: Earthscan.

Cook, J., Guttmann, E. B. A. and Mudd, A. (2004). Excavations of an Iron Age Site at Coxwell Road, Faringdon. *Oxoniensia* LXIX, 181–285.

Cook, H. and Williamson, T., (2007). *Water meadows: history, ecology and conservation*. Macclesfield: Windgather Press.

Cooney, G. (2000). *Landscapes of Neolithic Ireland*. London: Routledge

Crawford B. E. and Ballin Smith, B. (1999). The Biggings, Papa Stour, Shetland: The History and Excavation of a royal Norwegian Farm. Edinburgh: Society of Antiquaries of Scotland & Det Norske Videnskaps-Akademi, Vol. 15.

Cressey, G. B. (1958). Qanats, karez and foggaras. *Geographical Review* 48(1), 27–44.

Cunliffe, B. (1991). *Iron Age Communities in Britain* (3rd edition). London and New York: Routledge.

Dabaieh, M. (2013). Earth vernacular architecture in the Western Desert of Egypt. In M. Markku (ed.), *VERNADOC RWW 2002*, 24–30. Lund University.

Davis, I. (2006). Sheltering from extreme hazards. In L. Asquith and M. Vellinga (eds), *Vernacular Architecture in the Twenty-First Century: theory, education and practice*, 145–154. London: Taylor and Francis.

Deb, D. (2009). Valuing folk crop varieties for Agroecology and food security. *Independent Science News*, 26 October 2009. Available online at: http://independentsciencenews.org/un-sustainable-farming/valuing-folk-crop-varieties/ (accessed 14 November 2018).

Deccan Development Society (no date). Available online at: www.ddsindia.com (accessed 9 November 2018).

Deffeyes, K. S. (2001). *Hubbert's Peak: the impending world of oil shortage*. Princeton: Princeton University Press.

De Kraker, A. M. J. (2015). Flooding in river mouths: human caused or natural events? Five centuries of flooding events in the SW Netherlands, 1500–2000. *Hydrology and Earth System Sciences* 19, 2673–2684.

Denevan, W. M. (1992). The pristine myth: the landscape of the Americas in 1492. *Annals of the Association of American Geographers* 82, 369–385.

Denevan, W. M. (2001). *Cultivated Landscapes of Native Amazonia and the Andes*. Oxford: Oxford University Press.

Denevan, W. M., Mathewson, K. and Knapp, G. (eds) (1987). *Prehispanic Agricultural Terraces in the Andean Region*. Oxford: British Archaeological Reports, International Series 359 (i).

Dent, D. (2005). Overview of agrobiologicals and alternatives to synthetic pesticides. In J. Pretty (ed.), *The Pesticide Detox*, 70–82. London: Earthscan.

Derpsch, R. (2004). History of Crop Production With & Without Tillage. *Leading Edge, The Journal of No-Till Agriculture*, No-till On The Plains Inc., Wamego KS, March 2004, 3(1), 150–154.

Diamond, J. (2005). *Collapse: How societies choose to fail or survive*. London: Penguin.

Díaz, J. P. and Harris, P. (2008). Urban agriculture in Havana: opportunities for the future. In A.Viljoen, K. Bohn and J. Howe (eds), *Continuous Productive Urban Landscapes: designing urban agriculture for sustainable cities*, 136–145. Oxford and Burlington, MA: Architectural Press.

Dick, R. P., Sandorb, J. A. and Eash, N. S. (1994). Soil enzyme activities after 1500 years of terrace agriculture in the Colca Valley, Peru. *Agriculture, Ecosystems & Environment* 50(2), 123–131.

Dimakopoulos, S. (2016). Agricultural Terraces in Classical and Hellenistic Greece. In G. L. M. Burgers, S. J. Kluiving and R. A. E. Hermans (eds), *Multi-, Inter- and Transdisciplinary Research in Landscape*

Archaeology. Proceedings of the 3rd International Landscape Archaeology Conference LAC2014. Amsterdam: Vrije Universiteit Amsterdam.

Dockrill, S. J., Bond, J., Smith, A. N. and Nicholson, R. A. (2007). *Investigations in Sanday, Orkney, Vol. 2: Tofts Ness, Sanday – An island landscape through 3000 years of prehistory*. Kirkwall: The Orcadian Ltd.

Donkin, R. (1979). *Agricultural Terracing in the Aboriginal New World*. Tucson: University of Arizona Press.

Doolittle, W. E. (2000). *Cultivated Landscapes of Native North America*. Oxford: Oxford University Press.

Dutton, R. W. (1989). Aflaj renewal in Araqi: a village case study from Oman. In P. Beaumont, M. Bonine and K. Mclachalan (eds), *Qanat, Kariz and Khattara*, 237–256. Wisbech: Menas Press.

Eastwood, R. J. and Hughes, C. E. (2008). Origins of domestication of *Lupinus mutabilis* in the Andes. In *Lupins for Health and Wealth, Proceedings of the 12th International Lupin Conference, 2008*, 373–379. Freemantle, Western Australia: International Lupin Association.

Emery, V. L. (2011). Mud-Brick Architecture. In W. Wendrich (ed.), UCLA Encyclopedia of Egyptology, Los Angeles. Available online at: https://escholarship.org/content/qt4983w678/qt4983w678.pdf (accessed 14 November 18).

ENSSER (2013). The European Network of Scientists for Social and Environmental Responsibility. Available online at: https://ensser.org/press_release/pr01-15/ (accessed 14 November 18).

Erickson, C. (1987). *The Dating of Raised-field Agriculture in the Lake Titicaca Basin, Peru*. British Archaeological Reports, International Series 359, 373–384.

Erickson, C. (1988). Raised Field Agriculture in the Lake Titicaca Basin: Putting Ancient Andean Agriculture Back to Work. *Expedition* 30(3), 8–16.

Erickson, C. (1992). Prehistoric landscape management in the Andean Highlands: raised field agriculture and its environmental impact. *Population and Environment* 13, 285–300.

Erickson, C. L. (1995). Archaeological methods for the study of ancient landscapes of the Llanos de Mojos in the Bolivian Amazon. In Stahl, P. W. (ed.), *Archaeology in the Lowland American Tropics: Current analytical methods and applications*, 66–95. Cambridge: Cambridge University Press.

Erickson, C. L. (2003). Agricultural Landscapes as World Heritage: Raised Field Agriculture in Bolivia and Peru. In J.-M. Teutonico and F. Matero (eds), *Managing Change: Sustainable Approaches to the Conservation of the Built Environment*, 181–204. Los Angeles: Getty Conservation Institute.

Erickson, C. L. (2014). Amazonia: The Historical Ecology of a Domesticated Landscape. In S. Hecht, K. Morrison and C. Padoch (eds), *The Social Lives of Forests*, 199–214. Chicago: University of Chicago Press.

Erickson, C. and Candler, K. L. (1989). Raised fields and sustainable agriculture in the Lake Titicaca Basin of Peru. In J. O. Browder (ed.), *Fragile Lands of Latin America: Strategies for Sustainable Development*, 230–248. Boulder, CO: Westview Press.

Evans, A. (2009). *The Feeding of the Nine Billion: Global food security for the 21st century*. London: Royal Institute of International Affairs.

Evenari, M., Shanan, L. and Tadmor, N. (1971). *The Negev: The challenge of a desert*. Cambridge, MA: Harvard University Press.

FAO (1995). *Dimensions of Need: atlas of food and agriculture*. Available online at: http://www.fao.org/docrep/U8480E/U8480E0D.HTM#Restoring%20the%20land (accessed 14 November 2018).

FAO (2004). What is Agrobiodiversity? Food and Agriculture Organization of the United Nations Factsheet. Available online at: http://www.fao.org/docrep/007/y5609e/y5609e01.htm#bm1 (accessed 14 November 2018). Also available in *Building on Gender, Agrobiodiversity and Local Knowledge*, FAO, 2004.

FAO (2017a). *Food and agriculture: Driving action across the 2030 Agenda for Sustainable Development*. Available online at: http://www.fao.org/3/a-i7454e.pdf (accessed 2 May 2018).

FAO (2017b). Land-resource planning key to feeding growing world, preserving resources. Available online at: http://www.fao.org/land-water/news-archive/news-detail/en/c/1034507/ (accessed 14 November 2018).

FAO (1980). Report on the 1980 World Census of Agriculture, cited in V. Shiva (2001), *Yoked to Death: Globalisation and Corporate Control of Agriculture.* New Delhi: Research Foundation for Science, Technology, and Ecology.

Fardous, N., Taimeh, A. and Jitan, M. (2004). Indigenous water-harvesting systems in Jordan. In T. Oweis, A. Hachuma and A. Bruggeman (eds), *Indigenous Water-Harvesting Systems in West Asia and North Africa*, 42–60. Aleppo: ICARDA.

Farr, D. (2008). *Sustainable Urbanism: urban design with nature.* Hoboken NJ: John Wiley & Sons, Inc.

Faulkner, E. H. (1943). Plowman's Folly. Oklahoma: University of Oklahoma Press.

Fazio, M., Moffett, M. and Wodehouse, L. (2009). *A World History of Architecture.* London: Laurence King.

Fedoroff, N. V., Battisti, D. S., Beachy, R. N., Cooper, P. J. M., Fischhoff, D. A., Hodges, C. N., Knauf, V. C., Lobell, D., Mazur, B. J., Molden, D., Reynolds, M. P, Ronald, P. C., Rosegrant, M. W., Sanchez, P. A., Vonshak, A. and Zhu, J.-K. (2010). Radically rethinking agriculture for the 21st century. *Science* 327(5967), 833–834.

Finucane, B. C. (2009). Maize and Sociopolitical Complexity in the Ayacucho Valley, Peru. *Current Anthropology* 50(4), 535–545.

Fokkens, H. (1998). *Drowned Landscape: the occupation of the western part of the Frisian-Drentian Plateau, 4400 BC–AD 500.* Amersfoort: Rijksdienst voor het Oudheidkundig Bodemonderzoek.

Francis, C. A. (1986). *Multiple Cropping Systems.* New York: MacMillan.

French, C. A. I., Macklin, M. G. and Passmore, D. G. (1992). Archaeology and palaeochannels in the Lower Welland and Nene valleys: alluvial archaeology at the fen-edge, eastern England. In S. Needham and M. G. Macklin (eds), *Alluvial Archaeology in Britain*, 169–176. Oxford: Oxbow Books.

Gammon, C. (2009). Weed-whacking herbicide proves deadly to human cells. *Scientific American*, June 23 2009.

Garrison, N. and Horowitz, C. (2012). *Looking Up: how green roofs and cool roofs can reduce energy use, address climate change, and protect water resources in Southern California.* Natural Resources Defense Council (NRDC) Report June 2012:12-06-B.

Ghaemmaghami, P. S. and Mahmoudi, M. (2005). Wind tower: a natural cooling system in Iranian traditional architecture. International Conference 'Passive and Low Energy Cooling for the Built Environment', May 2005, Santorini, Greece. Available online at: http://www.inive.org/members_area/medias/pdf/Inive%5Cpalenc%5C2005%5CGhaemmaghami.pdf (accessed 14 November 2018).

Gilbertson, D., Hunt, C. and Gillmore, G. (2000). Success, longevity and failure of arid-land agriculture: Romano-Libyan floodwater farming in the Tripolitanian pre-desert. In G. Barker and D. Gilbertson (eds), *The Archaeology of Drylands: Living at the margin.* One World Archaeology Vol. 39. London: Routledge.

Gilbertson, D. D., Schwenninger, J.-L., Kemp, R. A. and Rhodes, E. J. (1999). Sand-drift and soil formation along an exposed North Atlantic coastline: 14,000 years of diverse geomorphological, climatic and human impacts. *Journal of Archaeological Science*, 26, 439–469.

Gillam, C. (2011). Super weeds pose growing threat to.US crops. *Reuters*, Sept. 19 2011. Available online at: http://www.reuters.com/article/2011/09/20/us-monsanto-superweeds-idUSTRE78J3TN20110920 (accessed 14 November 2018).

Ginsburg, C. D. (1871). *The Moabite Stone: a fac-simile of the original inscription: with an English translation and a historical and critical commentary* (2nd edition). London: Reeves and Turner.

Girardet, H. (1999). *Creating Sustainable Cities.* Totnes: Green Books Ltd.

Gladstone, N., Bainbridge, D. and Stein, T. M. (no date). Buried clay pot irrigation. Available online at: https://www.doc-developpement-durable.org/file/programmes-de-sensibilisations/forets-protection/buried_clay_pot_PACE.pdf (accessed 1 August 2017).

Glaser, B. (2007). Prehistorically modified soils of Central Amazonia: a model for sustainable agriculture in the twenty-first century. *Philosophical Transactions of the Royal Society B* 362, 187–196.

Gliessman, S. R. (2007). *Agroecology: the ecology of sustainable food systems*. Boca Raton, Florida: CRC Press.

Goldsmith, E. (2003). How to feed people under a regime of climate change. *World Affairs Journal* 7(3). Available online at: http://www.culturechange.org/how_to_Goldsmith.html (accessed 14 November 18).

Goldsmith, E. and Hildyard, N. (1984). Overview. In E. Goldsmith and N. Hildyard *The Social and Environmental Effects of Large Dams: Volume 1*, Wadebridge: Wadebridge Ecological Center.

Greenpeace (2005). Decentralising Power: an energy revolution for the 21st century. Available online at: http://www.cwp-ltd.com/wp-content/uploads/2012/03/Greenpeace-DE-paper.pdf (accessed 14 November 2018).

Gregory, R. A., Murphy, E. M., Church, M. J., Edwards, K. J., Guttmann, E. B. and Simpson, D. D. A. (2005). Verification of a Mesolithic occupation in the Western Isles of Scotland? *The Holocene* 15(7), 944–950.

Groves, L. and Hinton, R. (2004). *Inclusive Aid: changing power and relationships in international development*. Abingdon: Earthscan.

Gupta, A. K. (2004). Origins of agriculture and domestication of plants and animals linked to early Holocene climate amelioration. *Current Science* 87(1), 54–59.

Guttmann, E.B.A., Dockrill, S.J. and Simpson, I.A. (2004). Arable agriculture in prehistory: new evidence from soils in the Northern Isles. *Proceedings of the Society of Antiquaries of Scotland* 134, 53–64.

Guttmann, E. B. A. (2005). Midden cultivation in prehistoric Britain: arable crops in gardens. *World Archaeology* 37(2), 224–239.

Guttmann, E. B., Simpson, I. A. and Davidson, D. A. (2005). Manuring practices in antiquity: a review of the evidence. In M. Brickley and D. Smith (eds), *Fertile Ground: Papers in Honour of Susan Limbrey*, 68–76. Oxford: Oxbow Books.

Guttmann, E. B., Simpson, I. A., Davidson, D. A. and Dockrill, S. J. (2006). The management of arable land in prehistory: case studies from the Northern Isles of Scotland. *Geoarchaeology* 21(1), 61–92.

Guttmann-Bond, E. (2010). Sustainability out of the past: how archaeology can save the planet. *World Archaeology* 42(3). 355–366.

Hagmann, J. and Murwira, K. (1996). Indigenous soil and water conservation in southern Zimbabwe. In C. Reij, I. Scoones and C. Toulmin (eds), *Sustaining the Soil: indigenous soil and water conservation in Africa*, 97–106. London: Earthscan.

Hall, A. (2013). The North Sea Flood of 1953. *Environment and Society Portal, Arcadia* 2013, no. 5, Rachel Carson Center for Environment and Society.

Halpern, D. (2015). *Inside the Nudge Unit: how small changes can make a big difference*. London: W. H. Allen.

Halstead, P. (1990). Waste not, want not: traditional responses to crop failure in Greece. *Rural History* 1, 147–164.

Hamilton, R., Flavell, R. B. and Goldberg, R. B. (2005). Genetics Provides Opportunity to Feed World, Experts Say. Available online at: http://www.monsanto.co.uk/search/display.phtml?uid=9528 (accessed 5 March 2010).

Harlan, J. R. (2008). Our vanishing genetic resources. In J. Pretty (ed.), *Sustainable Agriculture and Food Vol 1: History of Food and Agriculture*, 159–167. London: Earthscan.

Hastorf, C. A. (2009). Rio Balsas most likely region for maize domestication. *Proceedings of the National Academy of Sciences* 106(13), 4957–4958.

Henderson, C. (2001). Turning the world upside down. In *Recipes Against Hunger: success stories for the future of agriculture.* Greenpeace. Available online at: http://www.greenpeace.org/international/Global/international/planet-2/report/2009/4/recipes-against-hunger-succe.pdf (accessed 14 November 2018).

Henn, P. and Henning, J. (2002). Urban agriculture and sustainable urban systems: a benefits assessment of the garden movement in Havana, Cuba. *International Journal of Environmental and Sustainable Development* 1(3), 202–209.

Herring, R. (2009). Persistent narratives: why is the 'Failure of Bt cotton in India' story still with us? AgBioForum: *The Journal of Agrobiotechnology Management & Economics* 15(2), 14–22.

Hillel, D. (1991). *Out of the Earth: civilization and the life of the soil.* Berkeley, CA: University of California Press.

Hillel, D. (1994). *Rivers of Eden: The struggle for water and the quest for peace in the Middle East.* Oxford: Oxford University Press.

Hillman, G. C. (1981). Reconstructing crop husbandry practices from charred remains of crops. In R. Mercer (ed.), *Farming Practice in British Prehistory,* 123–162. Edinburgh: Edinburgh University Press.

Hillman, G. C. (2000). The plant food economy of Abu Hureyra 1 and 2. In A. M. T. Moore, G. C. Hillman and A. J. Legge (eds), *Village on the Euphrates: from foraging to farming at Abu Hureyra,* 327–399. Oxford: Oxford University Press.

Holt-Giménez, E. (2006). Movimiento Campesino a Campesino: linking sustainable agriculture and social change. *Backgrounder* 12(1), 1–4.

Howard, E. (1902). Garden Cities of To-Morrow (being the second edition of 'To-Morrow: a Peaceful Path to Real Reform'). London: Swan Sonnenschein & Co. Available online at: https://www.gutenberg.org/files/46134/46134-h/46134-h.htm (accessed 14 November 2018).

Howe, J., Bohn, K. and Viljoen, A. (2008). Food in time: the history of English open urban space as a European example. In A. Viljoen, K. Bohn and J. Howe (eds), *Continuous Productive Urban Landscapes: designing urban agriculture for sustainable cities,* 95–107. Oxford and Burlington, MA: Architectural Press.

Hunter J. (2000). Pool, Sanday and a Sequence for the Orcadian Neolithic. In: Ritchie A, (ed.) *Neolithic Orkney in its European Context* (1st edition), pp. 117–125. Cambridge: McDonald Institute for Archaeological Research.

Hunter, J., Bond, J., Dockrill, S. J. and Smith, B. (eds). Excavations on Sanday, Orkney. Unpublished draft report.

Hussain, I., Abu-Rizaiza O. S., Habib, M. A. A. and Ashfaq, M. (2008). Revitalizing a traditional dryland water supply system: The Karezes in Afghanistan, Iran, Pakistan, and the Kingdom of Saudi Arabia. *Water International* 33(3), 333–349

IAASTD (2009). International Assessment of Agricultural Knowledge, Science and Technology for Development. Summary for decision makers of the Sub Saharan Africa (SSA) report. Washington D.C.: Island Press.

IFAD (2012). *Sustainable Smallholder Agriculture: feeding the world, protecting the planet,* International Fund for Agricultural Development, Rome. Available online at: https://webapps.ifad.org/members/gc/35/docs/GC35-Concept-note.pdf (accessed 14 November 2018).

IPCC (2001). Climate Change 2001: The Scientific Basis. Contribution of Working Group I to the Third Assessment Report of the. Intergovernmental Panel on Climate Change. In J. T. Houghton, Y. Ding, D. J. Griggs, M. Noguer, P. J. van der Linden, X. Dai, K. Maskell and C. A. Johnson (eds), *Climate Change 2001: The Scientific Basis.* Cambridge: Cambridge University Press. Available online at: https://www.ipcc.ch/ipccreports/tar/wg1/pdf/WGI_TAR_full_report.pdf (accessed 14 November 2018).

IPCC (2014). Climate Change 2014 Synthesis Report. Available online at: http://ar5-syr.ipcc.ch/ipcc/ipcc/resources/pdf/IPCC_SynthesisReport.pdf (accessed 14 November 2018).

IPCC Working Group (2007). Can the warming of the 20th century be explained by natural variability? Available online at: https://www.ipcc.ch/publications_and_data/ar4/wg1/en/faq-9-2.html (accessed 21 July 2017).

Issar, A. (2014). To calm troubled waters. *New Scientist*, 3 July 2014.

Issar, A. and Zohar, M. (2007). *Climate Change: Environment and History of the Near East* (2nd edition). Heidelberg: Springer.

Jackson, M. D., Mulcahy, S. R., Chen, H., Li, Y., Li, Q., Cappelletti, P. and Wenk, H. R. (2017). Al-tobermorite mineral cements produced through low-temperature water-rock reactions in Roman marine concrete. *American Mineralogist* 102(7), 1435–1450.

Jenssen, S. M. (2006). Traditional Rainwater Harvesting in Jordan: A qualitative study of Project Rainkeep (1994–1995). MA thesis, The University of Bergen. Available online at: http://bora.uib.no/handle/1956/9863 (accessed 14 November 2018).

Jim, C. Y. (2017). An archaeological and historical exploration of the origins of green roofs. *Urban Forestry & Urban Greening* 27, 32–42.

Jones, G. (1990). The application of present-day cereal processing studies to charred archaeobotanical remains. *Circaea* 6(2), 91–96.

Kaiser, J. (1996). Pests overwhelm Bt cotton crop. *Science* 273(5274): 423–430.

Karrou, M. and Boutfirass, M. (2004). Indigenous water harvesting techniques in Morocco. In T. Oweis, A. Hachum and A. Bruggeman (eds), *Indigenous Water-Harvesting Systems in West Asia and North Africa*, 61–76. Aleppo: ICARDA.

Kassam, A., Friedrich, T., Derpsch, R. and Kienzle, J. (2015). Overview of the Worldwide Spread of Conservation Agriculture. *The Journal of Field Actions* 8 (Field Actions Science Reports). Available online at: http://factsreports.revues.org/3966 (accessed 14 November 2018).

Kedar, Y. (1957). Water and Soil from the Desert: Some Ancient Agricultural Achievements in the Central Negev. *The Geographical Journal* 123(2), 179–187.

Kellert, S. R. and Wilson, E. O. (1993). *The Biophilia Hypothesis*. Washington D.C.: Island Press.

Khan, M. A., Ansari, R., Ali, H., Gul, B. and Nielsen, B. L. (2009). A potentially sustainable cattle feed alternative to maize for saline areas. *Agriculture, Ecosystems & Environment* 129(4), 542–546.

Kimoto, Y., Ishikawa, H., Nakamata, H., Kobori, I. (1991). Karez, Water Resources in Xinjian, China. *Bulletin of the Faculty of Bioresources*, Mie University Vol. 6, 109–151. Available online at: http://ci.nii.ac.jp/naid/110000506994/en (accessed 14 November 2018).

King, F. H. (2004). *Farmers of Forty Centuries: organic farming in China, Korea and Japan*. Mineola, N.Y.: Dover Publications. (First published in 1911 as *Farmers of Forty Centuries; or Permanent Agriculture in China, Korea and Japan*).

Kirbaş, B. and Hizli, N. (2016). Learning from Vernacular Architecture: ecological solutions in traditional Erzurum houses. *Procedia: Social and Behavioural Sciences* 216, 788–799.

Kohler-Schneider, M. (2002). Contents of a storage pit from late Bronze Age Stillfried, Austria: another record of the 'new' glume wheat. *Vegetation History and Archaeobotany* 12, 105–111.

Kuijt, I. and Goring-Morris, N. (2002). Foraging, farming and social complexity in the Pre-Pottery Neolithic of the Southern Levant: a review and synthesis. *Journal of World Prehistory* 16(4), 361–440.

Lang, C. and Stump, D. (2017). Geoarchaeological evidence for the construction, irrigation, cultivation and resilience of the 15th–18th century AD terraced landscape at Engaruka, Tanzania. *Quaternary Research* 88(3), 382–399.

Latin America Press (2011). UN calls for transgenic corn ban. Available online at: http://lapress.org/articles.asp?art=6410 (accessed 14 November 2018).

Lau, B., Ford, B. and Hongru, Z. (2014). The environmental performance of a traditional courtyard house in China. In W. Weber and S. Yannas (eds), *Lessons from Vernacular Architecture*, 99–110. London: Routledge.

Lavee, H., Poesen, J. and Yair, A. (1997). Evidence of high efficiency water-harvesting by ancient farmers in the Negev Desert, Israel. *Journal of Arid Environments* 35, 341–348.

Legge, A. and Rowley-Conwy, P. A., (2000). The exploitation of animals. In A. M. T. Moore, G. C. Hillman and A. J. Legge (eds), *Village on the Euphrates: from foraging to farming at Abu Hureyra*, 423–471. Oxford: Oxford University Press.

Lightfoot, D. (1994). Morphology and ecology of lithic-mulch agriculture. *Geographical Review* 84(2), 172–185.

Lightfoot, D. (1996). The nature, history, and distribution of lithic mulch agriculture: an ancient technique of dryland agriculture. *The Agricultural History Review* 44(2), 206–222. Available online at: http://www.jstor.org/stable/40275100 (accessed 14 November 2018).

Lightfoot, D. R. (2000). The origin and diffusion of qanats in Arabia: new evidence from the northern and southern Peninsula. *The Geographical Journal* 166(3), 215–226.

Little, B. and Morton, T. (2001). *Building with Earth in Scotland: innovative design and sustainability*. Scottish Executive Control Research Unit. Edinburgh: The Stationary Office.

Lockeretz, W. (ed.) (2007). *Organic Farming: An International History*, Wallingford and Cambridge, MA: CAB International.

Luo, S. and Han, C. (1990). Ecological Agriculture in China. In C. A. Edwards, L. Rattan, P. Madden, R. H. Miller and G. House (eds), *Sustainable agricultural systems*, 299–322. Ankeny, Iowa: Soil and Water Conservation Society.

Lynas, M. (2001). A message from Bangladesh. In *Recipes Against Hunger: success stories for the future of.* Greenpeace. Available online at: http://www.greenpeace.org/international/Global/international/planet-2/report/2009/4/recipes-against-hunger-succe.pdf (accessed 14 November 2018).

Lynas, M. (2008). *Six Degrees: our future on a hotter planet*. London: Harper Collins.

MacKerron, D. K. L., Duncan, J. M., Hillman, J. R., Mackay, G. R., Robinson, D. J., Trudgill, D. L. and Wheatley, R. J. (1999). Organic farming: science and belief. *Scottish Crops Research Institute, Annual Report*, 1998/99, 60–72.

Macphail, R. I. (1981). Soil and botanical studies of the 'dark earth'. In M. Jones and G. W. Dimbleby (eds), *The Environment of Man: The Iron Age to the Anglo-Saxon period,* 309–331. Oxford: British Archaeological Reports, British Series 87.

Mann, C. C. (2002). The real dirt on rainforest fertility. *Science* 297, 920–923.

Mann, M. E., Bradley, R. S. and Hughes, M. K. (1998). Global-scale temperature patterns and climate forcing over the past six centuries. *Nature* 392, 779–787.

Mann, M. E., Bradley, R. S. and Hughes, M. K. (1999). Northern hemisphere temperatures during the past millennium: inferences, uncertainties, and limitations. *Geophysical Research Letters* 26, 759–762.

Mann, M. E. and Kump, L. R. (2015). *Dire Predictions: understanding climate change. The visual guide to the findings of the IPCC* (2nd edition). New York: Penguin Random House.

Manuel, M., Lightfoot, D. and Fattahi, M. (2017). The sustainability of ancient water control techniques in Iran: an overview. *Water History* 10(1), 13–30.

Maslen, G. (2017). *Too Late: how we lost the battle with climate change*. London and Melbourne: Hardie Grant Books.

Masood, E (2001). Tewolde Berhan Egziabher interview. *New Scientist* January 20(2274). Available online at: http://www.gmwatch.org/index.php/news/archive/2001/538-tewolde-berhan-egziabher-interview-new-scientist (accessed 30 October 2013).

Matin, N., Forrester, J. and Ensor, J. (2018). What is equitable resilience? *World Development* 109, 197–205. Available online at: https://www.sciencedirect.com/science/article/pii/S0305750X18301396?via%3Dihub (accessed 14 November 2018).

May, J. (2010). *Handmade Houses & Other Buildings: the world of vernacular architecture.* London: Thames and Hudson.

Mazoyer, M. and Roudart, L. (2006). *A History of World Agriculture: from the Neolithic Age to the current crisis*. London: Earthscan.

McCorriston, J. and Hole, F. (1991). The ecology of seasonal stress and the origins of agriculture in the Near East. *American Anthropologist* 93(1), 46–69.

McGill, G., Oyedele, L. O. and McAllister, K. (2015). An investigation of indoor air quality, thermal comfort and sick building syndrome symptoms in UK energy efficient homes. *Smart and Sustainable Built Environment* 4(3), 329–348.

McLaren, F. S. (2000). Revising the wheat crops of Neolithic Britain. In A. S. Fairbairn (ed.), *Plants in Neolithic Britain and Beyond*, 91–100. Oxford: Oxbow Books.

Meggers, B. J. (1954). Environmental limitation on the development of culture. *American Anthropologist* 56(5), 801–824.

Meggers, B. J. (1957). Environment and culture in the Amazon Basin: an appraisal of the theory of environmental determinism. Studies in Human Ecology, *Social Science Monographs* 3, 71–89. Washington DC: Pan American Union.

Meggers, B. (1971). *Amazonia: Man and Culture in a Counterfeit Paradise*. Chicago: Aldine.

Millar, D., Ayariga, R. and Anamoh, B. (1996). 'Grandfather's way of doing': gender relations and the yaba-itgo system in Upper East Region, Ghana. In C. Reij, I. Scoones and C. Toulmin (eds), *Sustaining the Soil: indigenous soil and water conservation in Africa*, 117–125. London: Earthscan.

Mitsch, W. J. and Gosselink, J. G. (2015). *Wetlands* (5th edition). Hoboken: John Wiley and Sons.

Mohamed, Y. A. (1996). Drought and the need to change: the expansion of water harvesting in Central Darfur, Sudan. In C. Reij, I. Scoones and C. Toulmin (eds), *Sustaining the Soil: indigenous soil and water conservation in Africa*, 35–43. London: Earthscan.

Moinar, T. J., Kahn, P. C., Ford, T. M., Funk, C. J. and Funk, C. R. (2013). Tree crops, a permanent agriculture: concepts from the past for a sustainable future. *Resources* 2(4), 457–488.

Monsanto (2010). Cotton in India. Available online at: https://www.thehindu.com/opinion/op-ed/a-perfect-storm-in-the-cotton-field/article23357894.ece (accessed 14 November 2018).

Montgomery, D. R. (2007). *Dirt: The erosion of civilizations*. Berkeley, Los Angeles, London: University of California Press.

Moustafa, A. T. and Mukred, A. W. (2002). Protected agriculture earns more income from less water for terrace farmers in Yemen. *Caravan*, 16, 22–23.

Munn, N. D. (1992). The Cultural Anthropology of Time: A Critical Essay. *Annual Review of Anthropology* 21, 93–123.

Nakanishi, N. (2004). China official says GMO cotton developing super pest. *Reuters*, 28 May 2004.

New Scientist (2002). Thought for Food. *New Scientist* 174(2343), 34.

Norman, M. J. T., Pearson, C. J. and Searle, P. G. E. (1984). *The Ecology of Tropical Food Crops*. Cambridge: Cambridge University Press.

Oliver, P. (1987). *Dwellings: the house across the world*. Oxford: Phaidon Press.

Ouedraogo, M. and Kaboré, V. (1996). The Zaï: a traditional technique for the rehabilitation of degraded land in the Yatenga, Burkina Faso. In C. Reij, I. Scoones and C. Toulmin (eds), *Sustaining the Soil: indigenous soil and water conservation in Africa*, 80–84. London: Earthscan.

Outerbridge, T. (1987). The disappearing chinampas of Xochimilco. *The Ecologist* 17(2), 76–83.

Oweis, T., Hachum, A. and Bruggeman, A. (2004). *Indigenous Water-Harvesting Systems in West Asia and North Africa*. Aleppo: ICARDA.

Oweis, T., Hachum, A. and Bruggeman, A. (2004). The role of indigenous knowledge in improving present water-harvesting practices. In T. Oweis, A. Hachum and A. Bruggeman (eds), *Indigenous Water-Harvesting Systems in West Asia and North Africa*, 1–20. Aleppo: ICARDA.

Özkan, S. (2006). Traditional and vernacular architecture in the twenty-first century. In L. Asquith and M. Vellinga (eds), *Vernacular Architecture in the Twenty-First Century: theory, education and practice*, 97–109. London: Taylor and Francis.

Paganelli, A., Gnazzo, V., Acosta, H., López, S. L. and Carrasco, A. E. (2010). Glyphosate-Based Herbicides Produce Teratogenic Effects on Vertebrates by Impairing Retinoic Acid Signalling. *Chemical Research in Toxicology* 23(10), 1586–1595.

Palmer, E. H. (1871). *The Desert of the Exodus*. Cambridge: Cambridge University Press.

Pape, J. C. (1970). Plaggen soils in the Netherlands. *Geoderma*, 4, 229–255.

Park, B. J., Tsunetsugu, Y., Kasetani,T., Kagawa, T. and Miyazaki, Y. (2010). The physiological effects of Shinrin-yoku (taking in the forest atmosphere or forest bathing): evidence from field experiments in 24 forests across Japan. *Environmental Health and Preventive Medicine* 15(1), 18–26.

Parker, H. (1965). Feddersen Wierde and Vallhagar: a contrast in settlements. *Medieval Archaeology* 9(1), 1–10.

Parker Pearson, M., Sharples, N. and Mulville, J. (1996). Brochs and Iron Age society. *Antiquity* 70, 57–68.

Parrott, N. and Marsden, T. (2002). *The Real Green Revolution: organic and agroecological farming in the South*. London: Greenpeace Environmental Trust.

Parsons, J. R. (1991). Political implications of Prehispanic chinampa agriculture in the Valley of Mexico. In H. R. Harvey (ed.), *Land and Politics in the Valley of Mexico: a two thousand year perspective*, 17–41. Albuquerque: University of Mexico Press.

Parsons, J. R., Parsons, M. H., Popper, V. and Taft, M. (1985). Chinampa agriculture and Aztec urbanization in the Valley of Mexico. In I. Farrington (ed.), *Prehistoric Intensive Agriculture in the Tropics*, 49–96. Oxford: British Archaeological Reports, International Series 232(i).

Pearce, F. (2001). An ordinary miracle. *New Scientist*, 169(2276), 16–17.

Pearce, F. (2002). Green Harvest: Chemical-free farming is paying dividends for the world's poor. *New Scientist* 173(2331), 16.

Pearce, F. (2004). To calm troubled waters (interview with Arie Issar). *New Scientist* (2454), 45–47.

Perfecto, I. (1995). Sustainable agriculture embedded in a global sustainable future: agriculture in the United States and Cuba. In B. Bryant (ed.), *Environmental Justice: Issues, policies, and solutions*, 172–186. Washington DC: Island Press.

Piperno, D. R., Ranere, A. J., Holst, I., Iriarte, J. and Dickau, R. (2009). Starch grain and phytolith evidence for early ninth millennium B.P. maize from the Central Balsas River Valley, Mexico. *Proceedings of the National Academy of Sciences* 106(13), 5019–5024.

Popper, V. (2000). Investigating chinampa farming. *Back Dirt: Newsletter of the Institute of Archaeology, University of California, Los Angeles*, Fall/Winter 2000, 4–5.

Postel, S. (1996). *Dividing the Waters: food security, ecosystem health, and the new politics of scarcity*. Worldwatch Paper 132. Washington: Worldwatch Institute.

Postel, S. (2012). Drip irrigation expanding worldwide. Available online at: https://voices. nationalgeographic.org/2012/06/25/drip-irrigation-expanding-worldwide/ (accessed 12 October 2017).

Postgate, J. N. (1992). *Early Mesopotamia: society and economy at the dawn of history*. London and New York: Routledge.

Pretty, J. (1998a). *The Living Land: Agriculture, food and community regeneration in rural Europe*. London: Earthscan.

Pretty, J. (1998b). Feeding the world? *The Splice of Life*, 4(6). Available online at: http://ngin.tripod. com/article2.htm. (accessed 1 August 2017)

Pretty, J. (2007). *Sustainable Agriculture and Food* (4 volumes). London: Earthscan Publications Ltd.

Pretty, J. and Bharucha, Z. P. (2014). Sustainable intensification in agricultural systems. *Annals of Botany* 2014, 1–26.

Pretty, J. and Bharucha, Z. P. (2015). Integrated pest management for sustainable intensification of agriculture in Asia and Africa. *Insects* 6(1), 152–182.

Pretty, J. N. and Shah, P. (1997). Making soil and water conservation sustainable: from coercion and control to Partnerships and Participation. *Land Degradation and Development* 8, 39–58.

Price, D. T. (1991). The Mesolithic of Northern Europe. *Annual Review of Anthropology* 20, 211–233.

Price, S. and Nixon, L. (2005). Ancient Greek agricultural terraces: evidence from texts and archaeological survey. *American Journal of Archaeology* 109(4), 665–694.

Qing, L. (2010). Effect of forest bathing trips on human immune function. *Environmental Health and Preventative Medicine* 15(1), 9–17.

Quist, D. and Chapela, I. (2001). Transgenic DNA introgressed into traditional maize landraces in Oaxaca, Mexico. *Nature* 414(6863), 541–543.

Rackham, O. (1986). *The History of the Countryside*. London: JM Dent & Sons.

Rambeau, C. and Black, C. (2011). Palaeoenvironments of the southern Levant 5000 BP to present: linking the geological and archaeological records. In S. Mithen and E. Black (eds), *Water, Life and Civilisation: climate, environment and society in the Jordan Valley*, 94–104. Cambridge University Press.

Rashid, M. and Ara, D. R. (2015). Modernity in tradition: reflections on building design and technology in the Asian vernacular. *Frontiers of Architectural Research* 4, 46–55.

Reicosky, D. C. and Saxton, K. E. (2007). The benefits of no-tillage. In C. J. Baker, K. E. Saxton, W. R. Ritchie, W. C. T. Chamen, D. C. Reicosky, F. Ribiero, S. E. Justice and P. R. Hobbs (eds), *No-Tillage Seeding in Conservation Agriculture* (2nd edition), 11–20. Wallingford: CAB International and FAO. Available online at: http://www.fao.org/docrep/012/al298e/al298e00.htm (accessed 14 November 2018).

Reifenberg, A. (1955). *The Struggle Between Sown Land and Desert*. Jerusalem: Bialik Institute Press.

Reij, C., Tappan, G. and Smale, M. (2009). *Agroenvironmental Transformation in the Sahel: another kind of 'Green Revolution'*. International Food Policy Research Institute Discussion Paper 00914.

Reimer, P. J., Bard, E., Bayliss, A., Beck, J. W., Blackwell, P. G., Ramsey, C. B., Buck, C. E., Cheng, H., Edwards, R. L., Friedrich, M., Grootes, P. M., Guilderson, T. P., Haflidason, H., Hajdas, I., Hatté, C., Heaton, T. J., Hoffmann, D. L., Hogg, A. G., Hughen, K. A., Kaiser, K. F., Kromer, B., Manning, S. W., Niu, M., Reimer, R. W., Richards, D. A., Scott, E. M., Southon, J. R., Staff, R. A., Turney, C. S. M. and van der Plicht, J. (2013). IntCal13 and Marine13 Radiocarbon Age Calibration Curves 0–50,000 Years cal BP. *Radiocarbon* 55(4), 1869–1887. Available online at: https://www.cambridge.org/core/journals/radiocarbon/article/intcal13-and-marine13-radiocarbon-age-calibration-curves-050000-years-cal-bp/FB97C1341F452BD6A410C6FE4E28E090 (accessed 14 November 2018).

Reynolds, P. J. (1995). Rural life and farming. In M. J. Green (ed.), *The Celtic World*, 176–209. London: Routledge.

Ribeiro, S. (2004). The day the sun dies: contamination and resistance in Mexico. *Seedling*, July 2004, 4–10.

Richards, M. P., Schulting, R. J. and Hedges, R. E. M. (2003). Sharp shift in diet at onset of Neolithic. *Nature* 425, 366.

Richards, W. R. 1682. *Wallography, or, The Britton describ'd being a pleasant relation of a journey into Wales ...: and also many choice observables ... of that countrey and people/by W.R., a mighty lover of Welch travels*. London: Printed for Edward Caudell, bookseller in Bath, 1682.

Rick, T. C., Erlandson, J. M. and Vellanoweth, R. L. (2001). Paleocoastal marine fishing on the Pacific coast of the Americas: perspectives from Daisy Cave, California. *American Antiquity* 66(4), 595–613.

Rigg, T. and Bruce, J. A. (1923). The Maori gravel soils of Waimea West, Nelson, New Zealand. *Journal of the Polynesian Society* 32, 85–93.

Ritchie, A. (1995). *Prehistoric Orkney*. London: Batsford.

Robinson, E. (2008). Designing the perfect weed – Palmer amaranth. Delta Farm Press. Available online at: http://deltafarmpress.com/management/designing-perfect-weed-palmer-amaranth (accessed 14 November 2018).

Robinson, S., Black, S., Sellwood, B. and Valdes, P. J. (2011). A review of palaeoclimates and palaeoenvironments in the Levant and Eastern Mediterranean from 25,000 to 5,000 years BP: setting the environmental background for the evolution of human civilisation. In S. Mithen and E. Black (eds), *Water, Life and Civilisation: Climate, Environment and Society in the Jordan Valley*, 71–93. International Hydrology Series. Cambridge: Cambridge University Press

Rogers, A. (2011). *Late Roman Towns in Britain: rethinking change and decline*. Cambridge: Cambridge University Press.

Rosen, S. A. (2000). The decline of desert agriculture: a view from the classical period Negev. In G. Barker and D. Gilbertson (eds), *The Archaeology of Drylands: Living at the margin*, 45–62. London and New York: Routledge.

Rossett, P. M. (2000). Cuba: a successful case study of sustainable agriculture. In F. Magdoff, J. B. Foster and F. H. Buttel (eds), *Hungry for Profit: the agribusiness threat to farmers, food and the environment*, 203–213. New York: Monthly Review Press.

Royal Society (2009). *Reaping the benefits: Science and the sustainable intensification of global agriculture*. Royal Society Policy Document 11/09.October 2009. Available online at: https://royalsociety. org/~/media/Royal_Society_Content/policy/publications/2009/4294967719.pdf (accessed 14 November 2018).

Ruddell, E. (1995). Growing food for thought: A new model of site-specific research from Bolivia. *Grassroots Development*, 19(1), 18–26.

Ruddle, K. and Zhong, G. F. (1988). *Integrated agriculture-aquaculture in South China, the dike-pond system of the Zhujiang Delta*. Cambridge: Cambridge University Press.

Sahai, S. and Rahman, S. (2003). Performance of Bt cotton: data from first commercial crop. *Economic and Political Weekly* 38 (30) 3139–3141. Available online at: www.epw.in https://www.jstor.org/stable/4413822?seq=1#page_scan_tab_contents (accessed 14 November 2018).

Santayana, G. (1905). *The Life of Reason: Reason in Common Sense*. New York: Scribner.

Scaife, R. G. and Burrin, P. J. (1992). Archaeological inferences from alluvial sediments: some findings from southern England. In S. Needham and M. G. Macklin (eds), *Alluvial Archaeology in Britain*, 75–91. Oxford: Oxbow Books.

Scholz, M. (2013). Water quality improvement performance of geotextiles within permeable pavement systems: a critical review. *Water* 5, 462–479.

Schug, G. R. (2011). *Bioarchaeology and climate change: a view from South Asian prehistory*. Gainesville: University Press of Florida.

Schumacher, E. F. (1974). *Small is Beautiful: A Study of Economics as if People Mattered*. London: Sphere Books Ltd.

Scialabba, N. E. and Hattam, C. (eds) (2002). *Organic Agriculture, Environment and Food Security*. Rome: FAO.

Scott, K. (2009). Irrigation system can grow crops with salt water. Available online at: http://www. wired.co.uk/article/irrigation-system-can-grow-crops-with-salt-water (accessed 1 August 2017).

Selhub, E. M. and Logan, A. C. (2012). *Your Brain on Nature: the science of nature's influence on your health, happiness and vitality*. Mississauga, Ontario: John Wiley & Sons Canada Ltd.

Serpell, J. (1989). Pet-keeping and animal domestication: a reappraisal. In J. Clutton-Brock (ed.), *The Walking Larder: Patterns of domestication, pastoralism, and predation*. London: Unwin Hyman.

Shah, V. R. and Tayyibji, R. (2008). The Kashmir House, it's seismic adequacy and the question of social sustainability. The 14th World Conference on Earthquake Engineering, Oct. 12–17, 2008, Beijing, China.

Shennan, C., Pisani Gareau, T. and Sirrine, J. R. (2005). Agroecological approaches to pest management in the US. In J. Pretty (ed.), *The Pesticide Detox*, 193–211. London and Sterling, VA: Earthscan.

Shiva, V. (2004). The future of food: countering globalisation and recolonisation of Indian agriculture. *Futures* 36, 715–732.

Simpson, I. A. (1993). The chronology of anthropogenic soil formation in Orkney. *Scottish Geographical Magazine*, 109, 4–11.

Simpson, I. A., Guttmann, E. B., Cluett, J. and Shepherd, A. (2006). Characterising anthropic sediments in North European Neolithic settlements: an assessment from Skara Brae, Orkney. *Geoarchaeology* 21(3), 221–235.

Slicher van Bath, B. H. (1963). *The Agrarian History of Western Europe, AD 500–1850.* London: Arnold.

Smits, A. J. M., Cals, M. J. R. and Drost, H. J. (2001). Evolution of European river basin management. In H. J. Nijland and M. J. R. Cals (eds), *River Restoration in Europe: practical approaches.* Conference on River Resoration, 41–48. Wageningen: Institute for Inland Water Management and Waste Water Treatment/RIZA Lelystad.

Speed, G. (2014). *Towns in the Dark? Urban transformations from late Roman Britain to Anglo-Saxon England.* Oxford: Archaeopress.

Spek, T. (1992). The age of plaggen soils. An evaluation of dating methods for plaggen soils in the Netherlands and Northern Germany. In A. Verhoeve and J. A. J. Vervloet (eds), *The transformation of the European rural landscape: Methodological issues and agrarian change,* 72–91. Brussels: National Fund for Scientific Research.

Spencer, J. E. and Hale, G. A. (1961). The Origin, nature and distribution of agricultural terracing. *Pacific Viewpoint* 2, 1–40.

Spokas, K. A., Cantrell, K. B., Novak, J. M., Archer, D. W., Ippolito J. A., Collins, H. P., Boateng, A. A., Lima, I. M., Lamb, M. C., McAloon, A. J., Lentz, R. D. and Nichols K. A. (2012). Biochar: A Synthesis of Its Agronomic Impact beyond Carbon Sequestration. *Journal of Environmental Quality* 41, 973–989.

Steele, J. (1997). *An Architecture for People: the complete works of Hassan Fathy.* London: Thames and Hudson.

Steiner, C., Teixeira, W. and Zech, W. (2004). Slash and char – an alternative to slash and burn practiced in the Amazon Basin. In B. Glaser and W. Woods (eds), *Amazonian Dark Earths,* 182–193. Heidelberg: Springer.

Tarolli, P., Preti, F. and Nunzio, R. (2014). Terraced landscapes: From an old best practice to a potential hazard for soil degradation due to land abandonment. *Anthropocene* 6, 10–25.

Tenenbaum, D. J. (2009). Biochar: carbon mitigation from the ground up. *Environmental Health Perspective* 117(2), A70–A73.

Tesemma, T. (1991). Improvement of indigenous durum wheat landraces in Ethiopia. In J. Engels, J. G. Hawkes and M. Worede (eds), *Plant Genetic Resources of Ethiopia,* 287–295. Cambridge: Cambridge University Press.

Tilman, D. (1998). The greening of the green revolution. *Nature,* 396, 211–212.

Ulrich, R. S. (1979). Visual Landscapes and Psychological Well-Being. *Landscape Research* 4(1), 17–23.

UNCCD (2009). An Introduction to the United Nations Convention to Combat Desertification. Fact Sheet 1. Available online at: http://www.unccd.int/publicinfo/factsheets/pdf/Fact_Sheets/Fact_sheet_01eng.pdf (accessed 16 December 2009).

UNEP (United Nations Environment Programme) (1990). Global Assessment of Soil Degradation (GLASOD). Available online at: http://www.isric.org/projects/global-assessment-human-induced-soil-degradation-glasod (accessed 14 November 2018).

Vandermeer, J. (1989). *The Ecology of Intercropping.* Cambridge: Cambridge University Press.

Van de Noort, R. and O'Sullivan, A. (2006). *Rethinking Wetland Archaeology.* London: Duckworth.

Van der Veen, M. (1995). The identification of maslin crops. In H. Kroll and R. Pasternak (eds), *Res Archaeobotanicae,* 335–343. Kiel: Kiel Inst. für Vor- und Frühgeschichte.

van de Westeringh, W. (1988). Man-made soils in the Netherlands, especially in sandy areas ('plaggen' soils). In W. Groenman-van Waateringe and M. Robinson (eds), *Man made soils,* 5–19. Symposia of the Association for Environmental Archaeology No. 6, BAR International Series 410. Oxford: British Archaeological Reports.

Van Zeist, W. (1968). Prehistoric and early historic food plants in the Netherlands. *Palaeohistoria* 14, 41–173.

Varisco, D. M. (1991). The future of terrace farming in Yemen: A development dilemma. *Agriculture and Human Values*, 8(1–2): 166–172.

Viljoen, A., Bohn, K. and Howe, J. (2008). *Continuous Productive Urban Landscapes: Designing urban agriculture for sustainable cities*. Oxford and Burlington, MA: Architectural Press (Elsevier).

Viljoen, A. and Howe, J. (2008). Cuba: laboratory for urban agriculture. In A. Viljoen, K. Bohn and J. Howe (eds), *Continuous Productive Urban Landscapes: Designing urban agriculture for sustainable cities*. 147–191. Oxford and Burlington, MA: Architectural Press (Elsevier).

Wåhlin, L. (1997). The family cistern: 3000 years of household water collection in Jordan. In M. Sabour and K. S. Vikør (eds), *Ethnic Encounter and Culture Change: Papers from the Third Nordic Conference on Middle Eastern Studies Joensuu June 1995*, 233–249. Bergen/London: Nordic Research on the Middle East 3.

Wall, E. and Smit, B. (2005). Climate change adaptation in light of sustainable agriculture. *Journal of Sustainable Agriculture* 27(1), 113–123.

Wells, P. S. (1984). *Farms, Villages, and Cities: commerce and urban origins in Late Prehistoric Europe*. Ithaca, NY: Cornell University Press.

Wenhua, L. (1993). *Integrated Farming Systems in China: An overview*. Veröffentlichungen des Geobotanischen Institutes der ETH, Stiftung Rübel, Zurich, 113. Heft.

Werner, L. (1992). Cultivating the secrets of Aztec gardens. *Americas* 44, 6–16.

Wilken, G. C. (1985). A note of buoyancy and other dubious characteristics of the 'floating' chinampas of Mexico. In I. Farrington (ed.), *Prehistoric Intensive Agriculture in the Tropics,* 49–96. Oxford: British Archaeological Reports, International Series 232(i).

William, E. (2010). *The Welsh Cottage: building traditions of the rural poor, 1750–1900*. Aberystwyth: RCAHMW.

Williamson, T. (2007). 'Floating' in context: meadows in the long term. In H. Cook and T. Williamson (eds), *Water Meadows: History, Ecology and Conservation*, 35–51. Macclesfield: Windgather Press.

Williamson, T. and Cook, H. (2007). Introducing water meadows. In H. Cook and T. Williamson (eds), *Water Meadows: History, Ecology and Conservation*, 1–7. Macclesfield: Windgather Press.

Wilson, E. O. 1984. *Biophilia*. Cambridge, MA: Harvard University Press.

Winnett, F. V. and Reed, W. L. (1964). *The Excavations at Dibon (Dhībân) in Moab. Part I: The First Campaign, 1950–1951. Part II: The Second Campaign, 1952*. Boston: The Annual of the American Schools of Oriental Research 1964.

Wirth, C. J. (1997). The governmental response to environmental degradation in the Xochimilco Ecological Zone of Mexico City. Paper delivered to the Latin American Studies Association, Guadalajara, Mexico, April 17–19, 1997.

Wittwer, S., Youtai, Y., Han, S. and Lianzheng, W. (1987). *Feeding a Billion: Frontiers of Chinese Agriculture*. East Lansing: Michigan State University Press.

Woods, W. I. and McCann, J. M. (1999). The anthropogenic origin and persistence of Amazonian Dark Earths. In Yearbook Conference of Latin Americanist Geographers Vol. 25, 7–14.

Woolley, C. L. and Lawrence, T. E. (2003). *The Wilderness of Zin*. London: Stacey International. First published in the Palestine Exploration Fund Annual, 1914–1915.

World Meteorological Organization (2016). Globally Averaged CO_2 Levels Reach 400 parts per million in 2015. Available online at: https://public.wmo.int/en/media/press-release/globally-averaged-co2-levels-reach-400-parts-million-2015 (accessed 22 June 2017).

Wright, J. (2009). *Sustainable Agriculture and Food Security in an Era of Oil Scarcity: Lessons from Cuba*. London and Sterling, VA: Earthscan.

Wright, K. R., McEwan, G. and Wright, R. M. (2006). *Tipon: Water Engineering Masterpiece of the Inca Empire.* Reston, VA: ASCE Press.

Wulff, H. E. (1968). The qanats of Iran. *Scientific American,* April 1968: 94–105.

WWF (2016). Risk and resilience in a new era. *Living Planet Report* 2016. Gland, Switzerland: WWF International. Available online at: http://awsassets.panda.org/downloads/lpr_living_planet_report_2016.pdf (accessed 14 November 2018).

Xinhau News Agency (2002). GM cotton damaging the environment. Available online at: http://news.xinhuanet.com/english/2002-06/03/content_422594.htm (accessed 10 June 2002)

Yair, A. (1983). Hillslope hydrology water harvesting and areal distribution of some ancient agricultural systems in the northern Negev desert. *Journal of Arid Environments* 6, 283–301.

Yates, D. (1999). Bronze Age field systems in the Thames Valley. *Oxford Journal of Archaeology* 18, 157–170.

Yates, D. (2001). Bronze Age agricultural intensification in the Thames Valley and Estuary. In J. Brück (ed.), *Bronze Age Landscapes: tradition and transformation,* 65–82. Oxford: Oxbow Books.

Yellen, J. E., Brooks, A. S., Cornelissen, E., Mehlman, M. J. and Stewart, K. (1995). A middle stone age worked bone industry from Katanda, Upper Semliki Valley, Zaire. *Science* 268(5210), 553–556.

Zhao, J. H., Ho, P. and Azadi, H. (2011). Benefits of Bt cotton counterbalanced by secondary pests? Perceptions of ecological change in China. *Environmental Monitoring and Assessment* 173(1–4), 985–994.

Zurayk, Rami A. (1994). Rehabilitating the ancient terraced lands of Lebanon. *Journal of Soil and Water Conservation* 49(2), 106.